UNSTEADY HEAT AND MASS TRANSFER IN HELICAL TUBE BUNDLES

B. V. Dzyubenko
G. A. Dreitser
L. -V. A. Ashmantas

◉HEMISPHERE PUBLISHING CORPORATION
A member of the Taylor & Francis Group
New York   Washington   Philadelphia   London

UNSTEADY HEAT AND MASS TRANSFER IN HELICAL TUBE BUNDLES

Copyright © 1990 by Hemisphere Publishing Corporation. All rights reserved. Printed in the United States of America. Except as permitted under the United States Copyright Act of 1976, no part of this publication may be reproduced or distributed in any form or by any means, or stored in a data base or retrieval system, without the prior written permission of the publisher.

Originally published as nestatsionarnyy teplomassoobmen v puchkakh vitykh trub by Mashinostroyeniye, Moscow, 1988.

Translated by E. A. Zharkova.

1 2 3 4 5 6 7 8 9 0   BRBR   9 8 7 6 5 4 3 2 1 0

This book was set in Times Roman by Edwards Brothers, Inc.
The editor was Jerry A. Orvedahl.
Cover design by Reneé Winfield.
Printing and binding by Braun-Brumfield, Inc.

A CIP catalog record for this book is available from the British Library.

Library of Congress Cataloging-in-Publication Data

Dzyubenko, B. V.
  [Nestatsionarnyi teplomassoobmen v puchkakh vitykh trub. English]
  Unsteady heat and mass transfer in helical tube bundles /
B. V. Dzyubenko, G. A. Drietser, L.-V. A. Ashmantas.
      p.   cm.
  Translation of Nestatsionarnyi teplomassoobmen v puchkakh vitykh trub.
    1. Heat—Transmission. 2. Mass transfer. 3. Tubes—Fluid dynamics. I. Dreitser, Genrikh Aleksandrovich. II. Ashmantas, L. -V. A. III. Title.
QC320.D9813   1990
621.402′2—dc20   89-71654
                                                            CIP

ISBN 0-89116-884-2

CONTENTS

	PREFACE	ix
	NOMENCLATURE	xiii
CHAPTER ONE	STATEMENT OF THE PROBLEM AND SPECIFIC FEATURES OF UNSTEADY HEAT AND MASS TRANSFER PROCESSES IN HELICAL TUBE BUNDLES	1
1.1	Helical tube bundle heat exchangers	1
1.2	Statement of problems on unsteady heat and mass transfer using the homogeneized flow model	4
1.3	Specific features of unsteady heat transfer processes in helical tube bundles	18
1.4	Specific features of unsteady heat and mass transfer processes	31
CHAPTER TWO	METHODS OF EXPERIMENTAL STUDY OF HEAT CARRIER MIXING	41
2.1	Methods of experimental study of heat carrier mixing under unsteady and steady conditions	41

2.2	Experimental set-ups for studying heat carrier mixing	47
2.3	Automatic experiment control apparata and a measuring system for studying unsteady heat and mass transfer	52

CHAPTER THREE — THE VORTEX FLOW STRUCTURE AND PHYSICAL MECHANISM OF HEAT AND MASS TRANSFER ENHANCEMENT — 65

3.1	The steady-state flow structure in a helical tube bundle	65
3.2	The unsteady flow structure at heating and cooling	73
3.3	Specific features of developing unsteady temperature fields under uniform heating of helical tubes	78

CHAPTER FOUR — TRANSPORT PROPERTIES OF FLOW UNDER STEADY CONDITIONS — 83

4.1	Heat and mass transfer in straight helical tube bundles	83
4.2	Heat and mass transfer in twisted bundles of helical tubes	98
4.3	Heat transfer and hydraulic resistance in helical tube bundles at uniform heat supply	108
4.4	Local heat transfer and hydraulic resistance at nonuniform heat supply over the bundle cross-section	115

CHAPTER FIVE — TRANSPORT PROPERTIES OF FLOW AND METHODS OF CALCULATING UNSTEADY HEAT AND MASS TRANSFER — 121

5.1	Theoretical calculation methods of unsteady temperature fields in a helical tube bundle	121
5.2	Unsteady mixing involving a sharp change in heat power	130
5.3	Contribution of the Re number and rate of achieving an operating regime to unsteady heat and mass transfer	140
5.4	Unsteady heat and mass transfer in helical tube bundles at different Fr_M numbers. Generalization of the experimental data on increasing heat load	146

5.5	Generalization of the results of unsteady heat and mass transfer at decreasing heat load	152
5.6	Unsteady heat and mass transfer at varying heat carrier flow rate	156

CHAPTER SIX METHODS OF EXPERIMENTAL STUDY OF UNSTEADY HEAT TRANSFER IN HELICAL TUBE BUNDLES 165

6.1	Methods of experimental study of unsteady heat transfer	165
6.2	Design of experimental set-ups	175
6.3	Measuring systems and apparatuses	179
6.4	Experiment procedure	183
6.5	Data processing	184

CHAPTER SEVEN UNSTEADY HEAT TRANSFER IN HELICAL TUBE BUNDLES 189

7.1	General information	189
7.2	Results of experimental study of unsteady heat transfer in tubes	189
7.3	Unsteady heat transfer in helical tube bundles	198

CHAPTER EIGHT METHODS OF CALCULATING UNSTEADY HEAT AND MASS TRANSFER. SOME RECOMMENDATIONS 209

8.1	Calculation of unsteady heat transfer at uniform heat supply to a bundle	209
8.2	Calculation of unsteady heat transfer in helical tube bundles with regard to interchannel mixing	211

REFERENCES 217

INDEX 221

PREFACE

Heat exchangers used in aviation engineering must have less size and mass at a prescribed heat power and have power for heat carrier pumping. Therefore, a necessity arises to develop rational methods of enhancing heat transfer in different cross-section channels, as well as to develop appropriate designs of heating surfaces. These are target-oriented artificial flow turbulization in a wall zone only [19,20] due to tube rolling and to smooth transverse protrusions inside the tubes and transverse grooves outside the tubes, and flow swirling in oval helical tubes in longitudinal and crossflows [39] implemented by drawing round tubes through a draw plate which gives their assigned shape and twisting as well as a controlled boundary layer separation in the crossflow past a tube bundle [14].

The method of flow swirling in helical tubes in longitudinal flow enables one not only to substantially decrease the size and mass (metal consumption) of heat exchangers but also to improve interchannel mixing of the heat carrier in the intertube space, which provides a levelling of temperature nonuniformities in a cross section of a helical tube bundle at nonuniform heat release (heat supply) and at a lateral heat carrier flow into a heat exchanger. Because of their advantages, helical tube heat exchangers can be used in different branches of industry.

In-line and crossflow helical tube heat exchangers were considered in [39], which dealt with detailed studies of turbulent flow structure, heat transfer, hydraulic resistance and mixing of a heat carrier, and experimental methods and engineering calculations of heat and mass transfer, as well as with estimation of the efficiency of such heating surfaces as compared to smooth-tube heat exchange apparata.

This book considers unsteady heat and mass transfer in helical tube bundles, since transient processes occurring due to varying operational conditions and the start or stop of heat exchange devices are decisive factors in a number of cases.

At present, the interest shown in unsteady convective heat transfer in channels can be explained by unsteady thermal processes which play an essential role in modern power plants, heat exchangers, and production equipment, and by increased requirements for the calculation accuracy of these devices operating with a high heat flux density. Unsteady thermal processes in these devices are characterized by a high rate of variation of parameters and are controlling, in a number of cases. Calculations of unsteady thermal processes in power plants, heat exchangers, major equipment, and mains must be based on the results of fundamental studies of unsteady convective heat transfer processes. These studies are necessary in order to be able to devise reliable methods of calculating temperature fields and thermal stresses, cooling and heating of pipelines, mains, and elements of propulsion systems and power plants, as well as those of optimizing these processes, of calculating transient operational conditions of different heat exchangers, and of developing automatic control systems.

Unsteady heat transfer in round tubes was studied in previous works, the results of which were generalized in [24]. In complex geometry channels formed by helical tube bundles, the unsteady heat and mass transfer processes have a number of specific features. These features are, first of all, stipulated by a design of in-line flow helical tube bundles which are responsible for complex spatial flow in such bundles and require special experimental and calculation methods to be elaborated with reference to physically grounded flow models.

In this book, consideration is given to a homogeneized flow model when applied to unsteady heat and mass transfer processes. In this case, together with the motion, energy, continuity, and state equations for a heat carrier, the mathematical description of this model also involves the heat conduction equation for a "solid phase," (i.e., for helical tubes), which allows for the impact of thermal inertia of tubes on a process. At the same time, it is possible to determine temperature fields in a heat carrier and in a solid phase from calculations.

In this monograph, experimental heat and mass transfer coefficients (turbulent diffusion and heat transfer coefficients) are proposed to close a system of equations for the turbulent flow in helical tube bundles. Experimental methods and special experimental set-ups allowing for the specific features of measurement of rapidly varying parameters have been developed to determine these coefficients. The flow model and the methods of calculating unsteady and steady heat and mass transfer processes have been experimentally verified on these devices.

Also, procedures are proposed to generalize experimental data on unsteady heat and mass transfer in helical tube bundles at different types of unsteadiness;

sharp and smooth change in heat load at the start and stop of a heat exchanger and at a transition from one operating condition to another, as well as at varying heat carrier flowrates. For this, the similarity and dimensional theories have been used, proposing similarity numbers and means to take into account the specific features of unsteady heat and mass transfer in helical tube bundles. Criterial relations are specified to calculate the effective diffusion, heat transfer, and hydraulic resistance coefficients under steady and unsteady conditions and are recommended for thermal and hydraulic calculations of heat exchangers. Consideration is given to the methods of calculating helical tube heat exchangers with regard to interchannel mixing, which enables one to determine both averaged and local parameters using the homogeneized flow model. Theoretical and experimental works of the present authors and other investigators are also analyzed.

The Preface, Chapters 1 and 4, and Sections 3.1 and 8.2 were written by B. V. Dzyubenko; Chapters 2 and 5 and Sections 4.1 and 4.2 were written by B. V. Dzyubenko and L. -V. A. Ashmantas; Section 3.3 was written by L. -V. A. Ashmantas; and Chapters 6 and 7 and Sections 1.3, 3.2, and 8.1 were written by G. A. Dreitser.

The authors are very grateful to Professor N. I. Melik-Pashaev, who reviewed this book and gave many valuable pieces of advice on improving it.

NOMENCLATURE

a	thermal diffusivity
b	middle width of a jet, $b = 2r_{mid}$
c	specific heat capacity
c_p	specific heat capacity at constant pressure
d	maximum size of an oval profile of a tube
d_{shell}	bundle diameter, $d_{shell} = 2r_{shell}$
d_{eq}	equivalent diameter
D_t	effective diffusion coefficient
E	spectral function of kinetic turbulence energy
f	frequency
F	cross-sectional area of tubes in a bundle
F_f	area of the flow cross section of a bundle
F_x	projection of mass forces on the x-axis
G	air mass flowrate
G_i	axial heat carrier flowrate in a cell
G_{ij}	heat carrier flow in the transverse direction from the cell i to the cell j per unit length of the channel
g	acceleration of gravity
i	enthalpy
K	dimensionless effective diffusion coefficient
K_{uns}	parameter of unsteady heat conduction
K_G	parameter of hydrodynamic unsteadiness
K^*_{Tg}	parameter of thermal unsteadiness

K_α	ratio of unsteady heat transfer coefficient to its quasi-stationary value
K_ξ	ratio of unsteady hydraulic resistance coefficient to its quasi-stationary value
L	spatial integral turbulence scale
l	bundle length; mixing path
m	bundle porosity with respect to heat carrier; $m = F_f / F_\Sigma$
N	heat power; number of tubes in a bundle
q	heat flux density
q_w	wall heat flux density
q_v	density of internal heat sources
p	static pressure
P	tube pitch in a bundle
P_t	total pressure
Δp	pressure drop
R	autocorrelation coefficient
r	radial coordinate
r_0	diffusion source radius
r_{shell}	bundle radius
r_{mid}	middle jet radius
$r_{d.s.}$	diffusion source location radius
S	tube pitch twisting
S_{tw}	helical tube twisting pitch relative to the bundle axis
T	temperature
T_b	mean-mass temperature
t	time
u, v, w	components of the averaged velocity V in the orthogonal coordinate system
u_{mean}	mean flowrate velocity
u', v', w'	components of a pulsational velocity
Δu	velocity excess in the flow core
v_τ, v_r	tangential and radial velocity components in the cylindrical coordinate system
V	averaged velocity vector modulus
v_1	rms pulsational velocity
x	longitudinal coordinate; distance from a diffusion source
$\overline{y^2}$	mean-statistical squared displacement
Z	special Reynolds number
α	heat transfer coefficient
α_m	dimensionless heat transfer coefficient
β	dimensionless friction coefficient; volumetric expansion coefficient
γ	angle of twisting of helical tubes relative to the bundle axis
Γ	integral time turbulence scale
δ	wall layer thickness; wall thickness
δ^*	displacement thickness of a boundary layer

ϵ_q	turbulent thermal diffusivity
κ	relative effective diffusion coefficient
λ	thermal conductivity
μ	dynamic viscosity coefficient
$\bar{\mu}$	interchannel mixing coefficient
ν	kinematic viscosity coefficient
$\nu_{\text{eff}}, \lambda_{\text{eff}}$	effective viscosity coefficient and thermal conductivity, respectively
ξ	hydraulic resistance coefficient
Π	channel perimeter
ρ	density
σ	rms deviation
τ	time; time delay
τ_{xw}	axial component of wall shear stress
τ_{zw}	tangential component of wall shear stress
τ_{z0}	tangential component of shear stress on the channel axis
$\tau_{\Sigma w}$	total wall shear stress
φ	angular coordinate
χ	overall heat transfer coefficient
Fo	Fourier number
Fr_{M}	number characterizing the specific features of flow in a helical tube bundle
Ho	dimensionless time (time homogeneity number)
Le	Lewis number
Nu	Nusselt number
Pr	Prandtl number
Re	Reynolds number

SUBSCRIPTS

tw, twisted	twisted bundle
d.s.	diffusion source
shell	heat exchanger shell
qs	quasi-stationary
L	Lagrange flow
max	maximum
M	modified
uns	unsteady
f	flow; straight helical tube bundle
w	wall
rod	rod
mean	average quantity; mean-mass
T	turbulent; "solid phase"
tube	tube
cyl	cylindrical part of a ribbed rod

E	Euler flow
b	mean mass
d	determined in terms of d_{eq}
m	at a mean temperature in a wall layer
r	along the r-axis
x	along the x-axis
δ	determined in terms of the wall layer thickness
τ	tangential
1, 2	flow inside helical tubes and in the intertube space, respectively

CHAPTER
ONE

STATEMENT OF THE PROBLEM AND SPECIFIC FEATURES OF UNSTEADY HEAT AND MASS TRANSFER PROCESSES IN HELICAL TUBE BUNDLES

1.1 HELICAL TUBE BUNDLE HEAT EXCHANGERS

Many diverse designs of oval helical tube heat exchangers have been developed. In the in-line flow helical tube heat exchanger (Fig. 1.1) the tubes are located relative to each other, are in contact over a maximum size of the oval, and are fastened with straight round ends in the tube plates.* Such a tube arrangement provides substantial improvement of heat and mass transfer processes in the intertube space of a heat exchanger, and another important problem is being solved, namely, provision of vibration strength. Heat transfer enhancement in the intertube space of such a heat exchanger and inside the helical tubes [39] at optimum relative twisting tube pitches $S/d = 6-15$ enables one to decrease a heat exchanger volume by 1.5–2 times, as compared to a smooth-tube heat exchanger at prescribed heat power and pumping power of heat carriers. In this case, the mass of the apparatus and its metal consumption decrease. In such an apparatus all the helical tubes are twisted in one direction (either left, or right). At the boundary of spiral

---

*Dzyubenko, B. V. and Vilemas, Yu. V. A shell-and-tube heat exchanger. Author's Certificate No. 761820 (USSR). Bulletin of Inventions, 1980, No. 33, p. 194.

channels of such tubes there appears tangential discontinuity of the rotational velocity component, thus resulting in flow turbulization. In the wall layer of the tubes, the flow is swirled according to the solid-state law, and in the core the flow swirling is specified by the interaction of spiral flows past adjacent tubes. Since the wall layer flow is swirled to a greater extent than the flow core (maximum of the rotational and radial velocity components is at the external boundary of the wall layer), use of helical tubes leads to flow turbulization, first of all, in the wall layer [39].

Therefore, in the intertube space of such apparatuses heat transfer and hydraulic resistance at the twisting pitches $S/d \approx 12$ increase, to an equal degree, in the turbulent flow while in the transient flow region at $Re = 10^3 - 10^4$ heat transfer enhancement advances a hydraulic resistance increase [52].

The specific features of flow and heat transfer in in-line flow helical tube heat exchangers result in decreasing the mass and overall dimensions of heat exchangers, which allows these devices to be used in different branches of industry. It is, therefore, of interest to study not only steady but also unsteady heat and mass transfer processes in in-line flow helical tube heat exchangers which may be of significance to develop reliable designs operating under high heat flux density conditions.

The nature of flow inside oval helical tubes both in a heat exchanger (Fig. 1.1) and in a cross-flow helical tube heat exchanger provides noticeable heat transfer enhancement [39].

Cross-flow helical tube heat exchangers may also be mounted over the maximum size of the oval, which improves their vibration strength characteristics; but in this case heat transfer improvement and levelling of temperature nonuniformities of the tubes over their perimeter are attained only if helical tubes are located in a manner to form slot channels along the tube bundle with a width being equal to a half difference between maximum and minimum sizes of the oval. In this case, the close-packed tubes are in contact

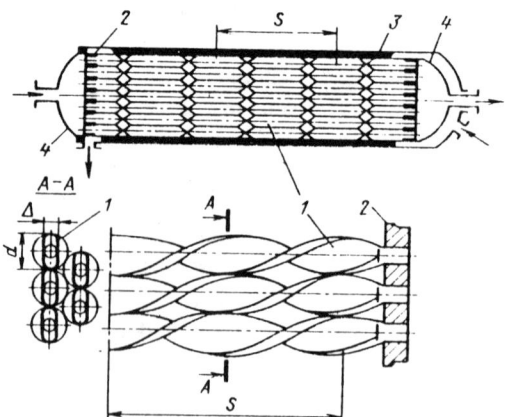

Figure 1.1 Swirled flow heat exchanger: *1)* helical tubes; *2)* tube plates; *3)* tube shell; *4)* bottoms.

STATEMENT OF THE PROBLEM 3

only with those of the adjacent rows.* The results on heat transfer, hydraulic resistance in such heat exchangers and the estimates of the efficiency of their use are presented in [39].

Another design of a cross-flow helical tube bundle heat exchanger, where spiral swirling of heat carrier in the intertube space causes both leveling of temperature nonuniformities over the tube perimeter and heat transfer enhancement, is characteristic of the cross arrangement of the adjacent rows of the helical tubes. In this case, a necessity arises to simultaneously heat or cool two different media. Additional flow turbulization in the intertube space is provided, in this case, by the interaction of different-direction helical flows due to vortex rotation at a transition of the flow from one tube row to another. Such a heat exchanger,† having two pairs of collectors with tube plates being at right angles located beneath the tubes of alternating rows, is characterized by a greater porosity of a bundle as against the previous one because the distance between the adjacent rows is increased $2/\sqrt{3}$ times at close packing of a bundle, and this provides each tube to be in contact with six coupled tubes over the twisting tube pitch. This heat exchanger is also more compact and requires less metal consumption as against the smooth tube one under the same heat power and pumping energy consumptions of heat carriers.

Twisting of helical tubes relative to the bundle axis‡ (Fig. 1.2) may be of certain interest for in-line flow heat exchangers at lateral inlet and outlet of heat carrier from the intertube space. In this case, it is possible to level nonuniformities of heat carrier velocity and temperature fields formed by the inlet conditions as well as by nonuniform heat supply over the bundle radius and azimuth due to azimuthal transfer of heat carrier by helical tubes twisted relative to the bundle axis. In this case, for better levelling of velocity and temperature field nonuniformities at the heat exchanger inlet and outlet, the connecting channels are formed for the medium of the intertube space. The porosity of these channels is greater than the one of the bundle because of use of straight ends of the tubes, whose diameter is equal to a smaller size of the oval. The results on heat transfer and hydraulic resistance in twisted helical tube bundles were examined in [39]. Revealed heat transfer enhancement in such bundles enables one to imply improvement of cross mixing of heat carrier, which is of importance when operating under non-

---

*Dzyubenko, B. V., Dreitser, G. A., Vilemas, Yu. V., Paramonov, N. V., Ashmantas, L.-V. A., Survila, Yu. V. A shell-and-tube heat exchanger. Author's Certificate No. 840662 (USSR), Bulletin of Inventions, 1981, No. 23, p. 178.

†Dzyubenko, B. V., Dreitser, G. A., Vilemas, Yu. V. et al. A shell-and-tube heat exchanger. Author's Certificate No. 1084583 (USSR). Bulletin of Inventions, 1984, No. 13, p. 127.

‡Dzyubenko, B. V., Vilemas, Yu. V., Varshkyavichyus et al. A shell-and-tube heat exchanger. Author's Certificate No. 937954 (USSR). Bulletin of Inventions, 1982, No. 23, p. 189.

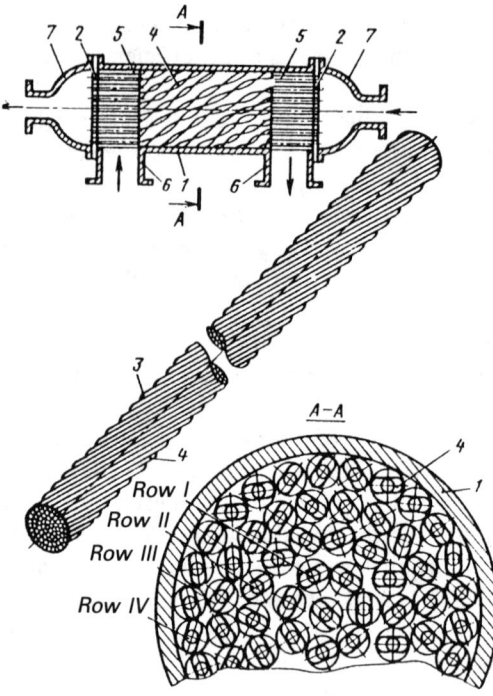

Figure 1.2 Heat exchanger with a helical tube twisted bundle: *1)* tube shell; *2)* tube plates; *3)* twisted bundle; *4)* helical tube; *5)* straight round ends of tubes; *6, 7)* connection pipes.

uniform heat supply over the radius and azimuth of the apparatus. In the present book this problem will be considered when generalizing the data on steady heat and mass transfer in straight helical tube bundles, in finned rod bundles and in bundles of spirally wound wire rods. In this case, the nature of flow is similar to the one in helical tube bundles.

1.2 STATEMENT OF PROBLEMS ON UNSTEADY HEAT AND MASS TRANSFER USING THE HOMOGENEIZED FLOW MODEL

The present monograph deals with problems on unsteady and steady heat and mass transfer when applied to heat exchange apparatuses and devices intended for aviation engineering. Heat and mass transfer processes in these apparatuses are enhanced by flow swirling in complex-geometry channels formed by oval spirally twisted helical tubes. The most complicated nature of flow is observed in the case of longitudinal flow past the bundles of helical tubes, both straight and twisted relative to the bundle axis.

In-line flow bundles of helical tubes of heat exchangers considered in Section 1.1 possess a number of the design features that cause complex

STATEMENT OF THE PROBLEM 5

spatial flow of heat carrier [3, 39]. Spatial flow nonuniformity in such bundles is attributed to the presence of alternating spiral and through channels, to tangential discontinuities, to the presence of the places, at which the adjacent tubes come in contact and behind which the flow is similar to the one in a wake behind a streamlined body. Secondary flow circulation due to centrifugal forces at spiral swirling of heat carrier by helical tubes substantially decreases a wall layer thickness on the tube walls and augments flow turbulence generated by a fixed wall. Thus, flow swirling to a greater extent turbulizes the wall layer, which improves heat transfer at moderate hydraulic losses. Appreciable anisotropy of the flow properties in a helical tube bundle is observed in a transition region of turbulent flow (Re = 10^3–$7 \cdot 10^3$). With increasing the Reynolds numbers for the developed turbulent flow region the flow core structure tends to a more isotropic structure. A transition from a laminar to a turbulent flow in a helical tube bundle occupies rather a wide region estimated by the range of the Reynolds numbers Re = $u d_{eq}/\nu \approx 10^2$–$10^3$. A swirling rate of the bundle flow is determined by a relative twisting pitch of helical tubes S/d.

In a general case, the unsteady flow of a homogeneous medium in helical tube bundles may be, on mathematical grounds, described by differential continuity equations [39]. The present work is concerned with turbulent flow. The differential equations describing this flow are derived from the system of the equations of Navier-Stokes, continuity and energy using the time-averaging rules at a fixed spatial point. The impact of the pulsational motion on the averaged one manifests itself, in this case, in increasing strain resistance in the averaged motion, and a necessity then arises to close a system of differential equations since the latter involve the mean correlated values of the products of the pulsational quantities: $\overline{u'v'} \neq 0$, $\overline{v'T'} \neq 0$, etc.

A system of the governing differential equations describing unsteady turbulent flow of homogeneous liquid with variable physical properties may be, in a tensor form, presented as [15]:

$$\rho \frac{\partial u_i}{\partial \tau} + \rho u_k \frac{\partial u_k}{\partial x_k} = -\frac{\partial p}{\partial x_i} + \frac{\partial}{\partial x_i}\left(\eta - \frac{2}{3}\mu\right)\frac{\partial u_l}{\partial x_l}$$

$$+ \frac{\partial}{\partial x_k}\left\{\mu\left(\frac{\partial u_i}{\partial x_k} + \frac{\partial u_k}{\partial x_i}\right) + \overline{\mu'\left(\frac{\partial u'_i}{\partial x_k} + \frac{\partial u'_k}{\partial x_i}\right)} - \overline{(\rho u_k)'u'_i}\right\}$$

$$\rho\frac{\partial I_0}{\partial \tau} + \rho u_k \frac{\partial I_0}{\partial x_k} = \frac{\partial p}{\partial \tau} + \frac{\partial}{\partial x_k}\left\{\mu u_i\left(\frac{\partial u_i}{\partial x_k} + \frac{\partial u_k}{\partial x_i}\right)\right.$$

$$+ \overline{\mu'\left[u_i\left(\frac{\partial u_i}{\partial x_k} + \frac{\partial u_k}{\partial x_i}\right)\right]'} + \left(\eta - \frac{2}{3}\mu\right)u_k\frac{\partial u_l}{\partial x_l} + \frac{\lambda}{c_p}\frac{\partial I}{\partial x_k}$$

$$\left. + \overline{\left(\frac{\lambda}{c_p}\right)'\frac{\partial I'}{\partial x_k}} - \overline{(\rho u_k)'I'_0}\right\}$$

$$\frac{\partial \rho}{\partial \tau} + \frac{\partial (\rho u_k)}{\partial x_k} = 0$$

$$I_0 = I + \frac{u^2}{2} + \frac{\overline{u'^2}}{2}$$

$$\rho, I, \mu, \eta, \lambda = f(p, T)$$

where u is the total velocity vector modulus, η the second viscosity coefficient, I the enthalpy, $I = c_p T$. A prime means the pulsational component of the quantity, and a bar stands for a mean value of the products of the pulsational quantities, $i = 1, 2, 3$ and $x_k's$ ($k = 1, 2, 3$) are the Cartesian coordinates.

Because of great mathematical difficulties it is impossible to obtain a general solution to the system of the differential equations describing unsteady turbulent motion in a helical tube bundle, considering that in solving unsteady heat transfer problems simultaneous analysis should be made both of the equations of motion of heat carrier and of those of heat conduction in a solid body, i.e., within the helical tube walls.

To make thermal and hydraulic calculations of such apparatuses, it is necessary to develop physically grounded flow models, to describe them mathematically using the systems of the differential equations which can be solved, to determine transfer coefficients needed to close a system of these equations, to study the effect of different parameters on heat and mass transfer coefficients and to establish criterial relations convenient for their calculation. Moreover, comprehensive studies must be made of the turbulent flow structure in such bundles, of velocity and temperature fields and of heat transfer and mixing under unsteady and steady conditions.

Calculation of unsteady heat transfer is associated with solving conjugated problems and involves difficulties that do not enable one, first of all, to obtain a closed system of equations for turbulent unsteady flow because there are no experimental data on the turbulent flow structure with a time-varying wall temperature. In [24], the methods were developed of studying unsteady heat transfer by solving the conjugated problems incorporating a one-dimensional description of processes in heat carrier. In addition, consideration is made of the equation of heat conduction of a channel wall:

$$\rho_w c_w \frac{\partial T_w}{\partial \tau} = \text{div}(\lambda_w \text{ grad } T_w) + q_v \tag{1.1}$$

and of the one-dimensional equations of motion, energy and continuity:

$$\frac{G}{u_f} \frac{\partial u_f}{\partial \tau} + G \frac{\partial u_f}{\partial x} = F_f \rho F_x - F_f \frac{\partial p}{\partial x} - \xi \frac{\rho u_f^2}{2 d_{eq}} F_f \tag{1.2}$$

$$\frac{G c_p}{u_f} \frac{\partial T_f}{\partial \tau} + G c_p \frac{\partial T_f}{\partial x} = \Pi q_w \tag{1.3}$$

$$\frac{\partial \rho}{\partial \tau} F_f + \frac{\partial G}{\partial x} = 0 \tag{1.4}$$

where

$$G = \rho u_f F_f \tag{1.5}$$

$$q_w = \alpha(T_w - T_f) \tag{1.6}$$

$$\xi = -\frac{\delta\left(\dfrac{\partial p}{\partial x}\right)}{\rho \dfrac{u_f^2}{2d_{eq}}} \tag{1.7}$$

where δ is the fraction of the longitudinal gradient $\partial p/\partial x$ spent for friction and velocity profile development.

System (1.1)–(1.7) is closed if the criterial relations for experimental α and ξ are known. In the case of unsteady heat transfer in tubes [24] it was marked that at a constant heat carrier flow rate a time variation of the wall temperature and heat flux affects the heat transfer coefficient because the turbulent flow structure changes, and unsteady heat conduction is imposed on quasi-stationary convective heat transfer. The time in the form of the numbers Fo $= a\tau/d^2$ or Ho $= u\tau/d$ explicitly does not enter into the criterial equations of unsteady heat transfer since the latter are obtained, assuming that unsteady heat transfer as against the quasi-stationary one is notable only for different temperature profiles in the wall layer $\beta d/2$ thick where β is the volumetric expansion coefficient. The above approach cannot be adopted to solve problems on unsteady mixing of heat carrier at nonuniform heat supply over the radius of a helical tube bundle as in this case either an axisymmetric or a three-dimensional problem on finding unsteady temperature fields in the bundle cross-section must be solved.

To determine unsteady temperature fields, it is advisable to use the homogeneized flow model as in the case of calculation of steady ones. The homogeneized flow model [39] consists in the following. A real bundle is replaced by a porous block, with its diameter being equal to the bundle one, in which a homogeneized medium or heat carrier is flowing with volumetric energy release (heat supply) sources q_v being distributed in them and with hydraulic resistance $\xi \rho u^2/2d_{eq}$, whose intensity varies over the bundle radius [9]. After the displacement thickness of the wall layer δ^* is determined and the material layer δ^* thick is conventionally built up on the walls of the tube, it is possible to consider free slip flow of a homogeneized medium, assuming that the velocity vector is parallel to the bundle axis and $\partial p/\partial r = 0$. Therefore, in the motion equation the velocity u represents the flow core one (outside the wall layer), the convective terms incorporating transverse velocity components are absent in LHS of the equation, and the diffusion term allows for the effect of different transfer mechanisms on velocity fields in cross-sections of a bundle [13]. Thus, the replacement of the real tube bundle flow by the homogenized one is an engineering tool to calculate velocity and temperature fields of heat carrier. The validity of this tool must be checked experimentally.

A system of differential equations incorporating those of motion, energy, continuity and state is solved to calculate temperature and velocity fields of heat carrier with its density as a function of temperature and pressure. For the axisymmetric problem this system is of the form:

$$\rho u \frac{\partial u}{\partial x} = -\frac{dp}{dx} + \frac{1}{r}\frac{\partial}{\partial r}\left(r\rho v_{\text{eff}}\frac{\partial u}{\partial r}\right) - \xi \frac{\rho u^2}{2d_{\text{eq}}} \quad (1.8)$$

$$G = 2\pi m \int_0^{r_{\text{sh}}} \rho u r \, dr \quad (1.9)$$

$$\rho u c_p \frac{\partial T}{\partial x} = q_v \frac{1-m}{m} + \frac{1}{r}\frac{\partial}{\partial r}\left(r\lambda_{\text{eff}}\frac{\partial T}{\partial r}\right) \quad (1.10)$$

$$p = \rho R T \quad (1.11)$$

In equation (1.10) the multiplier $(1 - m)/m$ is responsible for the homogenization effect. Boundary conditions of the problem are:

$$u(0, r) = u_{\text{in}} \quad T(0, r) = T_{\text{in}} \quad p(0, r) = p_{\text{in}} \quad (1.12)$$

$$\frac{\partial u(x, r)}{\partial r}\bigg|_{r=r_{\text{sh}}} = 0; \quad -\lambda \frac{\partial T(x, r)}{\partial r}\bigg|_{r=r_{\text{sh}}} = 0 \quad (1.13)$$

$$\frac{\partial u(x, r)}{\partial r}\bigg|_{r=0} = 0; \quad \frac{\partial T(x, r)}{\partial r}\bigg|_{r=0} = 0 \quad (1.14)$$

A system of nonlinear parabolic-type equations (1.8)–(1.12) with boundary conditions (1.12)–(1.14) may be solved by the network method using an explicit scheme, according to which the system of the equations is reduced to a dimensionless form and is written in finite differences. The form of the finite-difference analogs of the governing equations and the method of their solution when applied to the considered problem are presented in [9]. The solution algorithm has been implemented in the form of the BESM-4M computer program. Geometrical sizes of a bundle, heat carrier flow parameters at the bundle inlet, heat release (heat supply) q_v distributions over the bundle and radius as well as the physical properties of heat carrier are prescribed for calculation purposes.

To close the system of the equations, effective turbulent thermal conductivity λ_{eff}, viscosity coefficient v_{eff} and hydraulic resistance coefficient ξ are determined from experiment as functions of the similarity numbers characterizing the process [39].

That three-dimensional velocity and temperature fields of heat carrier be calculated, the system of the heat transfer and flow equations describing the flow of a homogenized medium in a helical tube bundle is of the form:

$$\rho u \frac{\partial u}{\partial x} = -\frac{dp}{dx} + \frac{1}{r}\frac{\partial}{\partial r}\left(r\rho v_{\text{eff}}\frac{\partial u}{\partial r}\right) + \frac{1}{r^2}\frac{\partial}{\partial \varphi}$$
$$\times \left(\rho v_{\text{eff}}\frac{du}{\partial \varphi}\right) - \xi\frac{\rho u^2}{2d_{\text{eq}}} \quad (1.15)$$

$$\rho u c_p \frac{\partial T}{\partial x} = q_v \frac{1-m}{m} + \frac{1}{r}\frac{\partial}{\partial r}\left(r\lambda_{\text{eff}}\frac{\partial}{\partial r}\right) + \frac{1}{r^2}$$
$$\times \frac{\partial}{\partial \varphi}\left(\lambda_{\text{eff}}\frac{\partial T}{\partial \varphi}\right) \quad (1.16)$$

$$G = m \int_0^{2\pi}\int_0^{r_{\text{sh}}} \rho u r \, dr \, d\varphi \quad (1.17)$$

$$p = \rho RT \quad (1.18)$$

In writing these equations, neglect was made of the convective terms for transverse and azimuthal velocity vector components in the equations of motion and energy as against the diffusional ones. Boundary conditions of the problem are:

$$u(r, \varphi, 0) = u_{\text{in}}(r, \varphi), \; T(r, \varphi, 0) = T_{\text{in}}(r, \varphi)$$
$$p(0) = p_{\text{in}} \quad (1.19)$$

$$\left.\frac{\partial u}{\partial r}\right|_{r=r_{\text{sh}}} = 0, \; \left.\frac{\partial T}{\partial r}\right|_{r=r_{\text{sh}}} = 0 \quad (1.20)$$

$$\left.\begin{array}{l} u(r, \varphi, x) = u(r, \varphi + 2\pi, x) \\ T(r, \varphi, x) = T(r, \varphi + 2\pi, x) \end{array}\right\} \quad (1.21)$$

The effective viscosity coefficient v_{eff} and thermal conductivity λ_{eff} in equations (1.8), (1.15), (1.10) and (1.16) take into account all the transfer mechanisms in a helical tube bundle: turbulent diffusion, convective transfer due to the vortex motion in bundle cells and ordered transfer via spiral channels of the tubes. The quantities v_{eff} and λ_{eff} are expressed in terms of the effective diffusion coefficient D_t, assuming that the turbulent Lewis and Prandtl numbers are equal to unity:

$$\text{Le}_T = \rho c_p D_t/\lambda_{\text{eff}} = 1 \quad (1.22)$$

$$\text{Pr}_T = \rho c_p v_{\text{eff}}/\lambda_{\text{eff}} = 1 \quad (1.23)$$

$$\lambda_{\text{eff}} = D_t \rho c_p \tag{1.24}$$

$$v_{\text{eff}} = D_t \tag{1.25}$$

The effective diffusion coefficient may be represented in a dimensionless form as

$$K = D_t/ud_{\text{eq}} \tag{1.26}$$

It is found experimentally and depends on the main similarity numbers. When determining the coefficient K in a helical tube bundle theoretical calculations were made at some prescribed values of this coefficient so that the design temperature fields of heat carrier formed a net, within which there were the measured values of a heat carrier temperature [16].

The system of equations (1.15)–(1.18) is solved both by the numerical methods when the numerical analogs of the equations are written according to the implicit scheme and by the matrix factorization method alongside with the iteration cycles by the nonlinearities [16]. In implementing this method, the greatest difficulty is associated with writing the finite-difference analogs of the governing equations at a singular point on the axis of a helical tube bundle ($r = 0$) and with including the periodicity conditions of unknown functions with respect to azimuth into one of the coefficient matrices.

To write numerical analogs of the governing equations for the bundle axis, use was made of Hershorin's method allowing one to express a value of the unknown function on the axis in terms of a set azimuthal values on the first design radius. To solve the finite-difference analogs of the energy and motion equations these are reduced to the form:

$$A_i \psi_{i+1} - B_i \psi_i + C_i \psi_{i-1} = -F_i \tag{1.27}$$

where ψ_i is the set of the azimuthal values of the unknown function on the i-th design radius. Moreover, the matrix A_i, B_i, C_i is composed of the coefficients of the equations. That the periodicity conditions be realized, the angle elements were introduced into the tridiagonal matrix B_i and allowed for these conditions on all the design radii of the domain of unknown functions. That the pressure gradient be omitted, the motion equation was preliminarily splitted into two equations using Simuni's substitution

$$u_{i,j} = w_{i,j} + z_{i,j}(dp/dx) \tag{1.28}$$

for each layer along the length, and the pressure gradient was determined from the integral relation for heat carrier flow rate over the cross-section of a helical tube bundle:

$$\frac{dp}{dx} = \frac{G - m \int_0^{2\pi} \int_0^{r_{\text{sh}}} w\rho r \, dr \, d\varphi}{m \int_0^{2\pi} \int_0^{r_{\text{sh}}} z\rho r \, dr \, d\varphi} \tag{1.29}$$

STATEMENT OF THE PROBLEM 11

The solution algorithm of the initial system of equations (1.15)–(1.18) under boundary conditions (1.19)–(1.21) was realized in the form of the FORTRAN program on the computer BESM-6. This program enables one to calculate values of a heat carrier temperature and velocity at 1500 nodes of the spatial network for 12–13 min when the thermophysical properties of heat carrier depend on flow parameters. This points to a high quick-response of the program.

In the case of the twisted bundle of helical tubes (Fig. 1.2), it is necessary to consider azimuthal heat and mass transfer by the twisted helical tubes relative to the bundle axis. Then, the system of the equations describing the steady-mean flow of a homogeneized medium in the intertube space of a heat exchanger with a twisted bundle of helical tubes will be of the form:

$$\rho u \frac{\partial u}{\partial x} + \frac{\rho v_\tau}{r} \frac{\partial u}{\partial \varphi} = -\frac{\partial p}{\partial x} - \frac{\xi \rho u}{2 d_{eq}} \sqrt{u^2 + v_\tau^2} \tag{1.30}$$

$$\rho u \frac{\partial v_\tau}{\partial x} + \frac{\rho v_\tau}{r} \frac{\partial v_\tau}{\partial \varphi} = -\frac{\xi_\varphi \rho v_\tau}{2 d_{eq}} \sqrt{u^2 + v_\tau^2} \tag{1.31}$$

$$\rho u c_p \frac{\partial T}{\partial x} + \frac{\rho v_\tau c_p}{r} \frac{\partial T}{\partial \varphi} = \frac{1}{r} \frac{\partial}{\partial r}\left(r \lambda_{eff} \frac{\partial T}{\partial r}\right) + \frac{1}{r^2}$$

$$\times \frac{\partial}{\partial \varphi}\left(\lambda_{eff} \frac{\partial T}{\partial \varphi}\right) + q_v \frac{1-m}{m} \tag{1.32}$$

This system must be supplemented with continuity (1.17) and state (1.18) equations. The main assumptions made in writing (1.30)–(1.32) implied a neglect of pressure drops in the bundle cross-section, which is consistent with the data [39], of intertube space heat losses via the heat exchanger shell and of the dissipative terms in motion equations (1.30) and (1.31). Boundary-value conditions of the problem are: at $x = 0$

$$u = u_{in},\ v_\tau = 0,\ T = T_{in},\ p = p_{in} \tag{1.33}$$

at

$$r = r_{sh} - \lambda \frac{\partial T}{\partial r} = 0,\ \frac{\partial u}{\partial r} = 0 \tag{1.34}$$

the periodicity conditions are:

$$u(r, \varphi, x) = u(r, \varphi, + 2\pi, x)$$

$$v_\tau(r, \varphi, x) = v_\tau(r, \varphi + 2\pi, x) \tag{1.35}$$

$$T(r, \varphi, x) = T(r, \varphi + 2\pi, x)$$

The system of equations (1.30)–(1.32), (1.17)–(1.18) may be solved numerically. In this case, the differential equations are replaced by their difference analogs using the generally accepted methods for an explicit scheme. This system of the equations is characteristic of neglecting the diffusional terms in the motion equations which are taken into consideration in a mathematical description of flow in straight helical tube bundles, (1.15)–(1.18). Therefore, that the system of equations (1.30)–(1.32), (1.17)–(1.18) be closed, there is no need to include the condition $\text{Pr}_T = 1$, and the quantity λ_{eff} related to the effective turbulent diffusion coefficient by (1.24) is determined from experiment.

When the homogenized flow model is applied to the unsteady process, alongside with the equations of motion, energy, continuity and state, it is necessary to consider an equation describing a temperature distribution in helical tubes (in a solid phase). In addition, temperature distributions of heat carrier and solid phase are determined, too. Thus, if the process is steady, then the one-temperature homogenized flow model for a real helical tube bundle (when only temperature fields of heat carrier are calculated) is used; if the process is unsteady, then the two-temperature model is adopted. Therefore, use of the homogenized flow model to calculate unsteady temperature fields in a helical tube bundle requires additional testing as such an approach can affect the thermal inertia properties of the homogenized flow model. On mathematical grounds, the problem on an axisymmetric non-uniform heat release field in the cross-section of a helical tube bundle may be represented by the following system of equations [27]:

$$\rho_{s.p} c_{s.p} \frac{\partial T_T}{\partial \tau} = q_v - \frac{4\alpha m}{(1-m)d_{eq}} (T_T - T)$$

$$+ \frac{1}{r} \frac{\partial}{\partial r} \left(r \lambda_{s.p} \frac{\partial T_T}{\partial r} \right) + \frac{\partial}{\partial x} \left(\lambda_{s.p} \frac{\partial T_T}{\partial x} \right) \tag{1.36}$$

$$\rho c_p \frac{\partial T}{\partial \tau} + \rho u c_p \frac{\partial T}{\partial x} = \frac{\partial p}{\partial \tau} + \frac{4\alpha}{d_{eq}} (T_T - T)$$

$$+ \frac{1}{r} \frac{\partial}{\partial r} \left(r \lambda_{\text{eff}} \frac{\partial T}{\partial r} \right) + \frac{\partial}{\partial x} \left(\lambda_{\text{eff}} \frac{\partial T}{\partial x} \right) \tag{1.37}$$

$$\rho \frac{\partial u}{\partial \tau} + \rho u \frac{\partial u}{\partial x} = -\frac{\partial p}{\partial x} - \xi \frac{\rho u^2}{2 d_{eq}} + \frac{1}{r} \frac{\partial}{\partial r} \left(r \nu_{\text{eff}} \frac{\partial u}{\partial r} \right) \tag{1.38}$$

$$\frac{\partial \rho}{\partial \tau} + \frac{\partial (\rho u)}{\partial x} = 0 \tag{1.39}$$

$$p = \rho R T \tag{1.40}$$

In obtaining the system of equations (1.36)–(1.40), an assumption was made that transverse velocity components were much less than the longitudinal one, $\partial p/\partial r = 0$, and the number M $\leq$ 0.5. Moreover, a neglect was made of heat and momentum transfer due to molecular diffusion, of heat release at kinetic energy dissipation and of longitudinal turbulent diffusion, and the porosity m was considered not to depend on the coordinates.

Single-valuedness conditions, i.e. the configuration and sizes of a helical tube bundle, physical conditions (kind of liquid, parameters characteristic of its physical properties), initial conditions (distribution of T, u and p at $\tau = 0$) and boundary conditions must be assigned to solve the system of equations (1.36)–(1.40).

In solving the system of equations (1.36)–(1.40), one prescribes the following geometrical sizes: tube bundle diameter, bundle length, flow cross-section of a bundle F_f and its porosity with respect to heat carrier, heated and wetted perimeters of a bundle, maximum size of the oval of a helical tube d, tube twisting pitch S and equivalent bundle diameter. Therefore, in solving the problem in the homogeneized flow statement, the volume of helical tubes of a bundle is determined to be such as in the case of a real tube bundle. This makes a conclusion that the thermal inertia properties of a real bundle and a solid phase of a homogeneized medium are identical if the physical properties of a solid phase and a real helical heat carrier-filled tube are identical. The physical properties of a solid phase are characterized by the density $\rho_{s.p.}$, heat capacity $c_{s.p.}$ and thermal conductivity $\lambda_{s.p.}$ (equation (1.36)). For each specific case these are chosen, assuming that the thermal inertia properties of a real bundle and homogeneized flow model are similar. This problem with reference to particular experimental conditions will be examined in Chapter 5.

The quantities λ_{eff}, ν_{eff}, α and ξ must be also determined experimentally to close the system of equations (1.36)–(1.40). If it is assumed that as in the case of the steady-state problem the turbulent numbers $Pr_T = 1$ and $Le_T = 1$ (expressions (1.22) and (1.23)), then it is necessary, from experiment, to determine values of heat transfer and hydraulic resistance coefficients α and ξ as well as of the effective diffusion coefficient D_t or of the dimensionless coefficient K according to (1.26) under unsteady conditions and their dependence on the similarity numbers characterizing a process. The coefficient K is determined by comparing the experimental and predicted temperature fields of heat carrier at each time instant τ.

Temperature fields may be calculated by solving numerically the system of equations (1.36)–(1.40) which should be supplemented with the following boundary conditions:

at the bundle inlet ($x = 0$)

$$T_T(r, 0, \tau) = T_{T.in}(r,\tau), \quad T(r, 0, \tau) = T_{in}(r, \tau)$$

$$p_f(0, \tau) = p_{f.in}(\tau) \tag{1.41}$$

at the bundle outlet (no heat transfer)

$$\frac{\partial T_T(r, x, \tau)}{\partial x}\bigg|_{x=l} = 0, \quad \frac{\partial T(r, x, \tau)}{\partial x}\bigg|_{x=l} = 0 \qquad (1.42)$$

$$p(x, \tau)\big|_{x=l} = p_{out}(\tau)$$

on the bundle axis (axial symmetry)

$$\frac{\partial T_T(r, x, \tau)}{\partial r}\bigg|_{r=0} = 0, \quad \frac{\partial T(r, x, \tau)}{\partial r}\bigg|_{r=0} = 0$$

$$\frac{\partial u}{\partial r}\bigg|_{r=0} = 0 \qquad (1.43)$$

at the external boundary of a bundle

$$-\lambda_T \frac{\partial T_T(r, x, \tau)}{\partial r}\bigg|_{r=r_{sh}} = 0, \quad -\lambda \frac{\partial T(r, x, \tau)}{\partial r}\bigg|_{r=r_{sh}} = 0$$

$$\frac{\partial u}{\partial r}\bigg|_{r=r_{sh}} = 0 \qquad (1.44)$$

Initial conditions are found by solving the steady-state problem at the time instant $\tau = 0$. In solving system (1.36)–(1.40), the quantities at the derivatives are preliminary averaged as a function of differentiation coordinates, removed from the differentiation sign and are then refined in the iteration cycles.

Heat transfer and energy equations may be solved by the variable-direction method [34]. Numerical analogs of these equations in this case are broken down according to the implicit scheme and are solved by the elimination method. A two-step explicit scheme [34, 35] may be adopted to solve motion and continuity equations. Thus, solving the problem is divided into two successive stages: solving the heat transfer equations and simultaneously solving the motion and continuity equations which are then coupled in terms of the state equation and iteration cycles.

In the next sections, the system of equations (1.36)–(1.40) will be simplified when applied to different types of unsteadiness, and an approach to solving problems on unsteady heat and mass transfer in helical tube bundles will be detailed. In addition, consideration will be also made of the experimental evidence of the accepted flow model, its mathematical description and the developed methods of solving the analyzed problems as well as of the problems on closing the systems of differential equations describing the flow of a homogenized medium. In these equations, the quantities v_{eff} and λ_{eff} expressed in terms of the coefficient D_t at $Le_T = 1$ and $Pr_T = 1$ will be determined empirically.

The adopted homogeneized flow model allows one to predict temperature and velocity fields in the cross-section of a bundle outside the wall layer. Then, considering the external flow to be prescribed, the boundary

STATEMENT OF THE PROBLEM **15**

layer approximations may be utilized to predict the flow in the wall layer. Within the framework of a mathematical description of the wall flow region where the viscosity forces exhibit, an assumption may be made that a pressure over the wall layer thickness is constant, and the equations for a flat boundary layer may be used. Such an approach is possible because of the small layer thickness δ as against the wall curvature radius of the spiral tube channel and small curvature of its spiral surface not undergoing sharp changes. In this case, the boundary layer equations under steady flow conditions are of the form [15]:

$$\rho u \frac{\partial u}{\partial x} + \rho v \frac{\partial u}{\partial y} = -\frac{dp}{dx} + \frac{\partial \tau}{\partial y} \tag{1.45}$$

$$\rho u c_p \frac{\partial T_0}{\partial x} + \rho v c_p \frac{\partial T_0}{\partial y} = \frac{\partial q_c}{\partial y} \tag{1.46}$$

$$\frac{\partial (\rho u)}{\partial x} + \frac{\partial (\rho v)}{\partial y} = 0 \tag{1.47}$$

where

$$T_0 = T + \frac{u^2}{2c_p} \tag{1.48}$$

$$\tau = \rho \nu \frac{\partial u}{\partial y} - \overline{\rho u'v'} \tag{1.49}$$

$$q_w = \lambda \frac{\partial T}{\partial y} - \overline{\rho v'T'} \tag{1.50}$$

$$\rho = \rho(p, T); \quad \nu = \nu(p, T); \quad \lambda = \lambda(p, T) \tag{1.51}$$

Turbulent shear stress and turbulent heat flux density may be given as:

$$-\overline{\rho u'v'} = \rho \nu_T \frac{\partial u}{\partial y}, \quad -\overline{\rho v'T'} = \lambda_T \frac{\partial T}{\partial y}$$

then

$$\tau = \rho (\nu + \nu_T) \frac{\partial u}{\partial y} \tag{1.52}$$

$$q = \rho c_p \left(\frac{\nu}{\text{Pr}} + \frac{\nu_T}{\text{Pr}_T} \right) \frac{\partial T}{\partial y} \tag{1.53}$$

Quantities ν_T and λ_T are determined either empirically or using the semi-empirical turbulence theories. When ν_T and λ_T are determined empirically, it is assumed that the turbulence characteristics at each flow point depend only on its local parameters at this point $\nu_T = \nu_T(\rho, \nu \partial u/\partial y, l)$, $\lambda_T = \lambda_T(\rho,$

v, $\partial u/\partial y$, l, λ) where l is the mixing length. Then, to determine l/δ and v_T/v, it is possible to use Nikuradse's formula

$$l/\delta = 0.14 - 0.08(1 - y/\delta)^2 - 0.06(1 - y/\delta)^4 \tag{1.54}$$

and Reichardt's formula

$$\frac{v_T}{v} = 0.4\left\{\eta - 7.15\left[\text{th}\left(\frac{\eta}{7.15}\right) + \tfrac{1}{3}\text{th}^3\left(\frac{\eta}{7.15}\right)\right]\right\} \tag{1.55}$$

For the flow of a variable-property liquid [15]

$$\eta = \frac{l\sqrt{\tau/\rho}}{0.4v} \tag{1.56}$$

The quantity v_T may be also determined by Prandtl's formula

$$v_T = l^2 \frac{\partial u}{\partial y} \tag{1.57}$$

where the mixing length l is found by van Driest's formula

$$l/y = 0.4[1 - \exp(\eta/26)], \text{ and } \eta = \frac{y\sqrt{\tau/\rho}}{v} \tag{1.58}$$

Then

$$\frac{v_T}{v} = \tfrac{1}{2}\{\sqrt{1 + 0.64\eta^2[1 - \exp(-\eta/26)]^2} - 1\} \tag{1.59}$$

Expressions (1.52)–(1.56) allow the system of equations (1.45)–(1.47) (in helical tube bundles filled with gas heat carrier $\text{Pr}_T = 1$) to be closed.

The semi-empirical expressions used to determine the quantities v_T and λ_T or $a_T = \lambda_T/(\rho c_p)$ were obtained in [15] by means of the turbulence balance equation for the intermediate boundary layer region where the turbulence energy production is approximately equal to dissipation:

$$\frac{v_T}{v} = -\frac{1-\alpha}{2\alpha} + \sqrt{\left(\frac{1-\alpha}{2\alpha}\right)^2 + \left(\frac{l\sqrt{\tau/\rho}}{v}\right)^2} \tag{1.60}$$

$$\frac{a_T}{v} = -\frac{1}{2\beta\text{Pr}} + \sqrt{\left(\frac{1}{2\beta\text{Pr}}\right)^2 + \frac{1}{\text{Pr}_T^2}\frac{v_T}{v}\left(\frac{v_T}{v} + \frac{1}{\alpha}\right)} \tag{1.61}$$

where α and β are the empirical constants, $\alpha, \beta < 1$. At $l\sqrt{\tau/\rho}/v \geq 1$, i.e. far from the wall, from expression (1.6), we have $v_T \approx l\sqrt{\tau/\rho}$, which is also given by formula (1.57) for turbulent friction. Equation (1.60) gives the best agreement with experiment at $\alpha = 0.15$ [15]. Besides the empirical

STATEMENT OF THE PROBLEM 17

constant β, equation (1.61) incorporates the extra empirical constant $\overline{Pr_T}$. Instead of equation (1.61), the expression

$$Pr_T = \frac{\nu_T/\nu}{-\frac{1}{2\beta Pr} + \sqrt{\left(\frac{1}{2\beta Pr}\right)^2 + \frac{1}{\overline{Pr_T^2}} \frac{\nu_T}{\nu}\left(\frac{\nu_T}{\nu} + \frac{1}{\alpha}\right)}} \quad (1.62)$$

may be used. Far from the wall ($\nu_T/\nu \gg 1$, or $\nu_T/\nu \gg 1/(2\beta Pr)$, from expression (1.62), we obtain $Pr_T \approx Pr$, i.e. the empirical constant $\overline{Pr_T}$ is the turbulent Prandtl number far from a wall.

A wall temperature distribution over the length and the radius of a helical tube heat exchanger may be determined using different boundary layer methods when the external flow is predetermined and is calculated by solving a system of equations describing the flow of a homogenized medium. These may be either the calculation methods or the ones, in which the initial system of two-dimensional equations is approximately replaced by that of one-dimensional ones. The latter methods are, in a number of cases more simple and convenient since these may be refined using the experimental data on heat transfer and hydraulic resistance coefficients, and on velocity and temperature fields. Such a method to calculate a boundary layer was elaborated in [15]. According to this method, the one-dimensional equations are being solved by the rapidly converging successive approximations. To close a system of equations, when a boundary layer is calculated by this method, Chapter 4 gives experimental grounds for relationships between the dimensionless parameters used to calculate heat transfer and hydraulic resistance at nonuniform heat supply using the homogenized flow model.

It should be noted that use of the local similarity principle of the turbulent transfer theory developed in the publications of a number of researchers, including V. M. Ievlev [15], has enabled one to extend the known semi-empirical turbulence theories to the case of flow in the wall layer of helical tubes. Conditions for use of this principle are determined from the analysis of the turbulence energy balance equation. The principal terms of this balance are those that are responsible for turbulence onset and suppression. In this case, the turbulent transfer characteristics at each point are determined through the averaged flow characteristics, volumetric force fields and turbulence properties l that enter into the turbulence energy balance equation. So, dimensionless relations (1.54)–(1.61) may be considered as universal local turbulent transfer laws. With increasing the scale l, the terms of the turbulence energy balance equation responsible for turbulence generation grow and the dissipative ones decrease. The quantity l may be, therefore, determined by the formula that incorporates geometrical flow parameters alone, e.g. (1.54).

In a helical tube bundle where there occurs flow swirling, the velocity and temperature fields in a wall layer are governed by the universal logarithmic laws which are formulated using the semi-empirical turbulence theories if as a characteristic length, use is made of the wall layer thickness determined by the relation

$$\delta = A d_{eq} Fr_M^{0.32} \tag{1.63}$$

where $A = 0.0349$ for δ_{max}, $A = 0.0156$ for δ_{min} [39], and the Fr_M number is a measure of the effect of flow swirling in a helical tube bundle. So, the experimental velocity fields in the bundle wall layer are described by a universal logarithmic profile

$$\frac{u}{\bar{u}} = 1 + \frac{\sqrt{\beta}}{0.39} \ln \frac{y}{\delta} \tag{1.64}$$

where

$$\beta = \tau_w / \rho \bar{u}^2 \tag{1.65}$$

Experimental temperature fields in the wall layer of helical tubes are also governed by a universal logarithmic profile [39]

$$\frac{T_w - T}{T_w - \bar{T}} = 1 + \frac{\alpha_m}{0.39 \sqrt{\beta}} \ln \frac{y}{\delta} \tag{1.66}$$

where

$$\alpha_m = \frac{q_w}{\rho \bar{u} c_p (T_w - \bar{T})} \tag{1.67}$$

A bar above $\bar{u}$ and $\bar{T}$ means that these are the parameters at the external boundary of the wall layer. In a helical tube bundle

$$\bar{u} = u_{mean} \left(1 - \frac{4\delta^*}{d_{eq}}\right)^{-1} \tag{1.68}$$

where δ^* is the displacement thickness of a boundary layer.

Thus, the experimental data evidence that the local similarity method for turbulent transfer of swirled flows in helical tube bundles may be adopted to calculate the fields u and T in a wall layer.

1.3 SPECIFIC FEATURES OF UNSTEADY HEAT TRANSFER PROCESSES IN HELICAL TUBE BUNDLES

The present section deals with a general statement of the problem on unsteady convective heat transfer in channels.

Development of the methods to calculate unsteady heat transfer from heat carrier in the turbulent channel flow is urgent for engineering purposes.

In a general case, such calculations are aimed at determining unsteady temperature and velocity fields in the heat carrier flow as well as temperature fields and thermal stresses in the material of a design surrounding the flow. In the majority of cases, it is enough to know mean-mass temperatures, mean flow rate velocity and pressure drops to describe the flow. In principle, these fields may be obtained by solving conjugated problems when the mathematical model for heat transfer and flow in heat carrier is supplemented with the heat conduction equation for the design material and with the conjugation conditions at the boundary between the heat carrier and the wall, and the boundary conditions are assigned at the external boundary of the channel walls [24].

However, solving the unsteady conjugated problems with reference to the majority of practically important cases faces still insuperable difficulties associated with making a great deal of calculations and with the impossibility to obtain a closed system of equations for turbulent unsteady flows even in the framework of the approximations of the semi-empirical turbulence theory because there are no experimental data on the turbulent flow structure at a time-varying channel wall temperature.

As the development of calculation procedures by solving three-dimensional conjugated problems offers these difficulties, it is most advisable to elaborate engineering methods by solving conjugated problems with the aid of a one-dimensional description of processes in heat carrier. Such an approach substantially simplifies the mathematical statement of the problem, making it quite solvable for numerical computations on modern computers and even in the form of nomograms. In this case, heat conduction equation (1.1) for channel walls is supplemented with one-dimensional equations of motion, energy and continuity (1.2)–(1.4).

Equation (1.3) is written, assuming that heat supply due to longitudinal heat leakages and energy dissipation due to friction may be ignored, as compared to that from the walls to heat carrier. System (1.1)–(1.4) may be closed if the equations for ξ and α are known. These equations may be, as a rule, obtained only from experiment.

Here a few remarks are in order. In calculating unsteady thermal processes, alongside with the concept of unsteady heat transfer and hydraulic resistance coefficients determined from (1.6) and (1.7), use is also made of those of quasi-stationary values of these coefficients. Use of the so-called quasi-stationary relations implies that for each moment of an unsteady process, heat transfer and hydraulic resistance are calculated by the formulas for a steady process. The values of the parameters of this process are equal to the instantaneous ones of unsteady process parameters at a considered time moment. Up to now the absence of reliable recommendations has impelled one to employ the quasi-stationary relations in practical calculations. As shown in [24], in many cases this provides prohibitive errors in practice. Therefore, in real unsteady processes it is of importance to know, how strongly non-stationary and quasi-stationary values of heat transfer and hydraulic resistance coefficients differ. The differences are specified by the relationships between these coefficients.

Let us analyze, what is responsible for deviations of unsteady heat transfer coefficients from their appropriate quasi-stationary values.

When the heat carrier flow rate in a channel (G = const) is constant, a time variation of the heat transfer coefficient α depends on that of the wall temperature T_w or heat flux density q_w. The time variation of T_w or q_w affects α due to changing turbulent flow structure and due to unsteady heat conduction imposed on quasi-stationary convective heat transfer. Theoretical studies made, as a rule, assuming the quasi-stationary flow structure, allow for a contribution of unsteady heat conduction alone. In this case, with heating a gas and with increasing a wall temperature ($\partial T_w/\partial \tau > 0$) the coefficient $K_\alpha = (\text{Nu}/\text{Nu}_{qs}) > 1$ (Nu and Nu_{qs} are the nonstationary and quasi-stationary values of the Nusselt number), and at $\partial T_w/\partial \tau < 0$ $K_\alpha < 1$. The variation of T_w affects the values of α due to re-developing temperature profile. Since the flow is turbulent, a temperature field variation in the flow core slightly influences α and its contribution is substantial only in the wall region. The wall heat flux propagates into the flow with a velocity proportional to $(a + \epsilon_q)/y$ (where a is the thermal diffusivity, ϵ_q is the turbulent thermal diffusivity and y is the distance from the wall). The estimates [24, 26] have shown that the time $\Delta \tau$, for which the heat impulse affects α, is rather small. Also, the distance along the flow axis, at which the heat impulse exerts influence, is small. Therefore, after the possible laws for $T_w(x, \tau)$ are expanded into the Taylor series, it may be shown that one may confine oneself only to the first linear expansion term. As a result, the dimensionless parameter is obtained and allows for the effect of the $T_w(\tau)$ variation on the Nusselt number [26]:

$$K_{T\tau} = \frac{\partial T_w}{\partial \tau} \cdot \frac{d_{eq}^2}{a(T_w - T_f)} \tag{1.69}$$

The heat flux contribution is taken into account by the parameter

$$K_{q\tau} = \frac{\partial q_w}{\partial \tau} \cdot \frac{d_{eq}^2}{q_w a} \tag{1.70}$$

Similarly, the effect of the variation of the wall temperature T_w along the channel is taken into consideration by the dimensionless parameter

$$K_{Tx} = \frac{\partial T_w}{\partial x} \cdot \frac{d_{eq}}{(T_w - T_f)} \tag{1.71}$$

and the contribution of the heat flux density varying along the channel, by the parameter

$$K_{qx} = \frac{\partial q_w}{\partial x} \cdot \frac{d_{eq}}{q_w} \qquad (1.72)$$

Numerical calculations of air steady heating at $Re_f = (1.6-2.3) \cdot 10^5$, $T_w/T_f = 1-2.2$ $K_{qx} = 0.01-0.013$ with regard to the variable properties of a gas and experiments with an electrically heated tube, whose wall thickness is variable, $d = 6.055$ mm and $l = 1081$ mm at $Re_f = (2.24-17.3) \cdot 10^4$, $T_w/T_f = 1.06-2.2$ $|K_{qx}| = 0.005-0.012$ have shown that for real values of K_{qx} its effect on heat transfer is not great [24]. In analyzing and generalizing the experimental data on unsteady heat transfer, this enables one not to allow for the effect of varying T_w and q_w along the channel or of the parameters K_{Tx} and K_{qx}.

Calculation of the contribution of unsteady heat conduction to heat transfer was made in the case of turbulent air flow at the hydrodynamic stabilization length, assuming the quasi-stationary turbulence structure and considering the variability of gas properties. A gas flow rate was considered constant, and $q_w(x, \tau)$ increased in time.

The energy equation

$$\rho \frac{\partial i}{\partial \tau} + \rho u \frac{\partial i}{\partial x} = \frac{1}{r} \cdot \frac{\partial}{\partial r} [(\lambda + \rho c_p \epsilon_q) \frac{\partial T}{\partial r}] \qquad (1.73)$$

was solved numerically by the polynomial approximations of the radius heat flux density distribution, which were before-hand supported by calculations

$$q = (\lambda + \rho c_p \epsilon_q) \frac{\partial T}{\partial r} = q_w(a_0 + a_1 R + a_2 R^2 + a_3 R^3) \qquad (1.74)$$

The turbulent flow structure was calculated by Reichardt's formula. To take into account the variability of gas properties the dimensionless distance from the wall $\eta = \sqrt{\xi/32}\, Re_w$ was determined through the values of ρ and μ at T_w. Calculation provided the convergence of the mean-mass enthalphy found by integration and obtained by solving the one-dimensional equation. It was shown that in virtue of high thermal diffusivity of a gas the impact of unsteady heat conduction is inconsiderable and is greatly less than the one revealed from the experimental data (Fig. 1.3). Similar results were obtained by solving numerically this problem with the aid of the finite-difference method at $Re_f = 10^4-3 \cdot 10^4$ on the computer BESM-6. In virtue of more low thermal diffusivity this impact is of more significance for liquids, however, the experimental data also witness a discrepancy with the predicted ones (Fig. 1.4) [24].

The insignificant effect of unsteady heat conduction on turbulent heat transfer of gases was also supported by the experimental data on varying the gas pressure and on the same mass flow rate and wall heat release (Chapter 7, [26]). In this case, the thermal diffusivity of a gas varied. It was found that at $G = $ const the heat transfer coefficient under unsteady conditions did

22 UNSTEADY HEAT AND MASS TRANSFER IN HELICAL TUBE BUNDLES

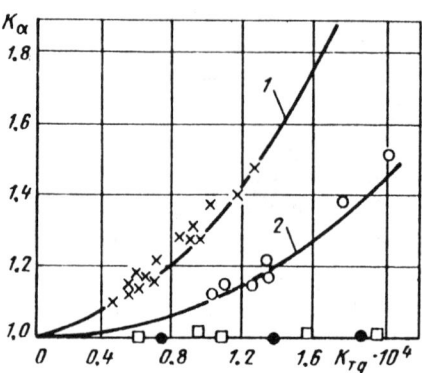

Figure 1.3 Comparison of the experimental and predicted data on unsteady heat transfer with air heating and with increasing heat load ($T_w/T_f = 1.1$): *1, 2*) experimental data at $\text{Re}_f = 10^5$ and $4.4 \cdot 10^5$; ■, ○, represents points obtained using the quasi-stationary turbulence model for $\text{Re}_f = (0.9–1.1) \cdot 10^5$ and $\text{Re}_f = (4.3–4.5) \cdot 10^5$ ($T_w/T_f = 1.08–1.12$); ×, ○ represents points obtained by the unsteady turbulence model for $\text{Re}_f = (0.9–1.1) \cdot 10^5$ and $(4.3–4.5) \cdot 10^5$ ($T_w/T_f = 1.08–1.12$). Here the thermal unsteadiness parameter $K_{Tg} = (\partial T_w/\partial \tau)(d_{eq}/(T_w - T_f)_o)\sqrt{\lambda/c_p g G}$ is found in terms of the temperature head ($T_w - T_f)_o$ in the final steady process.

not depend on a gas pressure (as in the case of the steady conditions), i.e. a substantial change in the gas thermal diffusivity at an invariable Re number did not exert an influence on unsteady heat transfer. So, it is left over to assume that in the case of turbulent gas flow the main reason for a difference

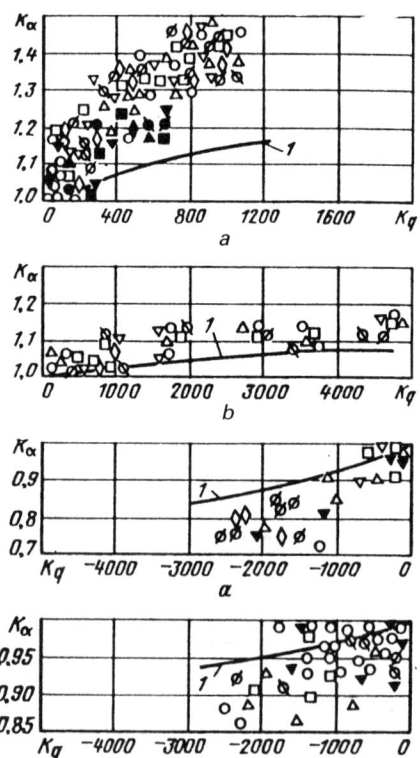

Figure 1.4 Comparison of the experimental and predicted data on unsteady heat transfer with water heating and with varying heat load: ○, □, △, ▽, ◇, ⊘, ○, experimental data for $x/d = 12.7, 36, 50, 82, 105, 128, 151$ at $\text{Pr}_f/\text{Pr}_w = 1–2$, respectively; ●, ■, ▲, ▼, ◆, ●, ●, the same data for x/d at $\text{Pr}_f/\text{Pr}_w = 2–3$; *a*) $\text{Re}_f = (7.5–10) \cdot 10^3$, $\text{Pr}_f = 8–12$; *b*) $\text{Re}_f = (5–7) \cdot 10^4$, $\text{Pr}_f = 4–6$; *c*) $\text{Re}_f = (3–4) \cdot 10^4$, $\text{Pr}_f = 6–8$; *1*) design curves.

between unsteady and quasi-stationary heat transfer is associated with the effect of unsteadiness on the turbulent flow structure. It is obvious that the thermal unsteadiness parameters used for generalizing the experimental data must be formed, considering this fact. In particular, unlike relations (1.69) and (1.70), these parameters must not depend on a gas pressure.

The analysis [24] is based on the studies of the turbulence generation mechanism. It is shown that the turbulence nature varies with a distance from a wall. In a viscous sublayer ($0 \leq \eta \leq 5$ where η is the dimensionless distance from a wall) the flow is not laminar. Small-amplitude velocity pulsations and a great amount of liquid from the adjacent regions penetrate into this sublayer. In the range $5 \leq \eta \leq 15$ there periodically appear vortex structures which are thrown into more remote layers. These ejections interact with the main flow primarily within the range $7 \leq \eta \leq 30$ and generate the layer turbulence, not exceeding $\eta = 70$. The onset and ejection of vortex structures from this layer lack any regular pattern in time and space and are affected by the local conditions, however, their intensity and frequency are the functions of performance parameters of averaged flow. A mean frequency of ejections is

$$\bar{\omega} = \omega_0 \, \text{Re}^{1.75} \tag{1.75}$$

where $\omega_0 = 10^{-7}$ 1/s. The flow core turbulence at $\eta > 70$ is transferred by convection and diffusion from the wall region and is characterized by a less intensity of velocity pulsations and by a large pulsation scale.

Analyzing this mechanism, it may be concluded that under unsteady conditions, of decisive role is, apparently, a local flow temperature variation over the range $\eta = 5-30$ for a mean time between vortex structures following each other at a given point.

According to (1.73) this time may be estimated as

$$\Delta \tau^* = \frac{1}{\bar{\omega}} = \frac{1}{\omega_0 \text{Re}^{1.75}} \tag{1.76}$$

It is rather small, and the value of the wall heating for the time $\Delta \tau^*$ may be, therefore, estimated by the linear temperature increment

$$\Delta T^* = \frac{\partial T_w}{\partial \tau} \Delta \tau^* \tag{1.77}$$

After the liquid is ejected and entrained, near the wall there appears a local region of decelerated liquid flow $\eta \leq 30$ thick with a very small velocity gradient. Then, this region of locally decelerated liquid flow interacts with the liquid bulk moving with the velocity close to the mean one for a given layer. As a result of this interaction, the liquid is sharply ejected from the decelerated flow region into the upper layers. This ejection is the main source of turbulent energy.

It may be assumed that under unsteady heating of a gas at $\partial T_w / \partial \tau > 0$, because of the decelerated heat carrier motion, the gas mass near the wall succeeds in heating and expanding to a great extent. This enlarges the surface of this mass interaction with large hot gas masses that fastly move relative to the cold gas and causes more intensive ejection of the cold gas. Therefore, on the one side, turbulence generation is enhanced and, on the other, ejection of some amount of the hot gas into the cold one is provided. However, as far as T_w/T_f is increased, the effect of varying the thermophysical properties and, especially a gas density near the wall will be potentiated, as compared to isothermal flow.

In the case of a postulated mechanism for the impact of unsteady wall heating on turbulence generation, it may be expected that the unsteadiness effect will be the greater, the larger is the volumetric expansion coefficient of a gas near the wall:

$$\beta_w = -\frac{1}{\rho}\left(\frac{\partial \rho}{\partial T}\right)_p \qquad (1.78)$$

A density variation for the heating time is

$$\Delta\rho = \rho_0 - \rho = \rho\beta(T - T_0) \approx \rho\beta_w \frac{\partial T_w}{\partial \tau}\Delta\tau^* \qquad (1.79)$$

Analysis of the impacts of thermal unsteadiness on a turbulent gas flow structure shows that a criterion for these impacts may be called the dimensionless parameter

$$K_{Tg} = \frac{\Delta\rho}{\rho} = \frac{\partial T_w}{\partial \tau} \cdot \frac{\beta_w}{\omega_0 Re^{1.75}} \qquad (1.80)$$

If account is taken of the fact that the Re effect is allowed for separately and the dimensionless constant ω_0 is replaced by the time scale $d_{eq}\sqrt{\lambda/c_p gG}$ [24], then the thermal unsteadiness parameter assumes the form:

$$K_{Tg} = \beta_w d_w \frac{\partial T_w}{\partial \tau}\sqrt{\frac{\lambda}{c_p gG}} \qquad (1.81)$$

For noncircular channels it is more convenient to use the time scale $\sqrt{\lambda/c_p g\rho u_f}$ and the dimensionless thermal unsteadiness parameter

$$K^*_{Tg} = \beta_w \frac{\partial T_w}{\partial \tau} \cdot \sqrt{\frac{\lambda}{c_p g\rho u_f}} \qquad (1.82)$$

Here λ, c_p and ρ are the thermal conductivity, heat capacity and density of heat carrier, $g = 9.8$ m/s^2, G is the mass flow rate, u_f is the mean mass flow velocity in the considered channel cross-section.

As already mentioned, if the influence of unsteady heat conduction on turbulent heat transfer is not substantial, which is typical of gas flow, then

STATEMENT OF THE PROBLEM 25

the unsteady heat transfer coefficient does not depend on a gas pressure. Therefore, the appropriate thermal unsteadiness parameter allowing for the effect of varying a turbulent flow structure on heat transfer should not be also affected by a gas pressure. Therefore, the constant d_{eq}^2/a (entering into expressions (1.69) and (1.70)) varying, for gases, proportional to a pressure, may not be used as a time scale in relation (1.80).

This was taken into account in obtaining the time scales $d_{eq}\sqrt{\lambda/c_p gG}$ and $\sqrt{\lambda/c_p g\rho u_f}$ being independent on a gas pressure.

The dimensional theory enables one also to obtain other dimensionless parameters which, in their turn, take into consideration the effect of varying a wall temperature on the turbulent flow structure and, in terms of it, on unsteady heat transfer:

$$K_{Tg}^{**} = \frac{\partial T_w}{\partial \tau} \beta_w \sqrt{\frac{d_{eq}}{g}} \tag{1.83}$$

$$K_{Tg}^{*'} = \beta_w d_{eq} \frac{\partial T_w}{\partial \tau} \sqrt{\frac{\mu}{gG}} \tag{1.84}$$

$$K_{Tg}'' = \beta_w \frac{\partial T_w}{\partial \tau} \sqrt{\frac{\mu}{g\rho u_f}} \tag{1.85}$$

where μ is the dynamic viscosity coefficient. As $G = u_f \rho F_f$ and $d_{eq} = 4F_f/\Pi$

$$K_{Tg}' = \frac{4\sqrt{F_f}}{\Pi} K_{Tg}^* \tag{1.86}$$

Parameters (1.81)–(1.85) are related in terms of the Re and Pr numbers:

$$K_{Tg}^* = K_{Tg}^{**} \sqrt{\frac{1}{\text{Re Pr}}} = K_{Tg}'' \sqrt{\frac{1}{\text{Pr}}} \tag{1.87}$$

$$K_{Tg}' = K_{Tg}^{*'} \sqrt{\frac{1}{\text{Pr}}} = K_{Tg}^{**} \frac{4}{\Pi} \sqrt{\frac{F_f}{\text{RePr}}}$$

$$= K_{Tg}'' \frac{4}{\Pi} \sqrt{\frac{F_f}{\text{Pr}}} \tag{1.88}$$

For gases at a wall temperature $\beta_w = 1/T_w$ and

$$K_{Tg}' = \frac{\partial T_w}{\partial \tau} \frac{d_{eq}}{T_w} \sqrt{\frac{\lambda}{c_p gG}} = d_{eq} \frac{\partial \ln T_w}{\partial \tau} \sqrt{\frac{\lambda}{c_p gG}}$$

$$K_{Tg}^* = \frac{\partial T_w}{\partial \tau} \frac{1}{T_w} \sqrt{\frac{\lambda}{c_p g\rho u_f}} = \frac{\partial \ln T_w}{\partial \tau} \sqrt{\frac{\lambda}{c_p g\rho u_f}}$$

$$K^{**}_{Tg} = \frac{\partial T_w}{\partial \tau} \frac{1}{T_w} \sqrt{\frac{d_{eq}}{g}} = \frac{\partial \ln T_w}{\partial \tau} \sqrt{\frac{d_{eq}}{g}}$$

$$K^{*\prime}_{Tg} = d_{eq} \frac{\partial T_w}{\partial \tau} \frac{1}{T_w} \sqrt{\frac{\mu}{gG}} = d_{eq} \frac{\partial \ln T_w}{\partial \tau} \sqrt{\frac{\mu}{gG}}$$

$$K^{\prime\prime}_{Tg} = \frac{\partial T_w}{\partial \tau} \frac{1}{T_w} \sqrt{\frac{\mu}{g\rho u_f}} = \frac{\partial \ln T_w}{\partial \tau} \sqrt{\frac{\mu}{g\rho u_f}}$$

The functional relation for Nu must be supplemented with one of the parameters (1.81)–(1.85) (for example, K^*_{Tg} by formula (1.82)) to take into account the contribution of the turbulent flow structure to unsteady heat transfer. In the case of unsteady nonisothermal flow the same parameter is also included into a relation for hydraulic resistance.

Since the functional relations for Nu and ξ must incorporate the parameters considering the flow nonisothermity, it is of no principal significance, through what temperature the values of thermophysical parameters entering into relations (1.81)–(1.85) are found. It is convenient to determine μ, λ, c_p and ρ entering into expressions (1.81)–(1.85) in terms of the mean-mass flow temperature T_f in the considered channel cross-section. As it follows from the above analysis, the value of the volumetric expansion coefficient β_w is determined through the wall temperature T_w.

In making the above-mentioned analysis it has been assumed that the heat carrier flow rate in a channel under unsteady conditions remains constant. At disturbances with respect to the heat carrier flow rate $G = G(\tau)$ its effect on heat transfer may mainly manifest itself in turbulent flow structure variations due to flow acceleration or deceleration.

As shown in [26], at time flow acceleration the shear stress and turbulent viscosity near the wall is greater than their quasi-stationary values (Fig. 1.5). And, vice versa, in the flow core the shear stress is less than its quasi-stationary value. This is possible only when the intensities of turbulent fluctuations and turbulent viscosity coefficient ϵ_σ decrease in the flow core. The overcome of the delay in the intensity of turbulent flow core fluctuations against their quasi-stationary values may be attained due to more intensive (as compared to quasi-stationary) diffusion of kinematic turbulent energy from the wall region. For this, the generation of kinetic turbulent energy near a wall should exceed a quasi-stationary value. This is possible if the velocity gradient, shear stress and turbulent viscosity near the wall are greater than their quasi-stationary values.

Thus, in the case of flow acceleration the turbulence intensity increases near the wall but decreases in the flow core. In this case, the turbulence flow core intensity may be less not only than its quasi-stationary value but also than its initial value before the flow is accelerated. Even the laminarization of the flow core is possible. And, vice versa, when the flow is de-

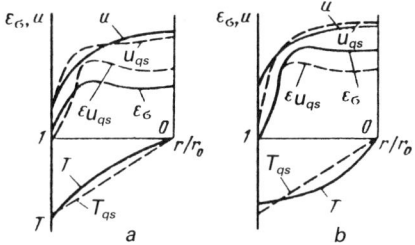

Figure 1.5 Qualitative time variations of shear stress profiles τ_x, kinematic turbulent viscosity coefficient ϵ_σ and velocity u for accelerated (a) and decelerated (b) flows (T_{qs}, ϵ_{uqs} and u_{qs} are the quasi-stationary values).

celerated, the turbulence intensity decreases near the wall and in the flow core. As shown in [24], in the case of developed turbulent flow (Re > (1.5–2) · 10⁴), heat transfer is mainly affected by varying the turbulence intensity near the wall. Thus, the flow acceleration increases heat transfer as against the quasi-stationary one and the flow deceleration decreases. With decreasing the Reynolds numbers (at Re < (1–1.5) · 10⁴) and, accordingly, with decreasing the flow turbulence intensity the effect of varying flow core turbulence proves to be predominant. The flow acceleration in this case may reduce heat transfer while the flow deceleration increases it.

In [24, 26] it is emphasized that in the case of turbulent flow the impact of a variable flow rate of heat carrier on heat transfer and hydraulic resistance is specified not by the law for $G(\tau)$ itself but only by its first derivatives $dG(\tau)/d\tau$ or by the dimensionless parameters of the form

$$K_{G1} = \frac{dG(\tau)}{d\tau} \frac{\nu}{G(\tau) u_f^2} \tag{1.89}$$

$$K_G = \frac{dG}{d\tau} \frac{d_{eq}^2}{\nu G} \tag{1.90}$$

$$K_u = \frac{\partial u_f}{\partial \tau} \frac{d_{eq}^2}{u_f \nu} \tag{1.91}$$

$$K_{u1} = \frac{\partial u_f}{\partial \tau} \frac{d_{eq}}{u_f^2} \tag{1.92}$$

$$K_{Gg} = \frac{dG}{d\tau} \frac{d_{eq}}{G} \sqrt{\frac{\mu}{gG}} \tag{1.93}$$

$$K_{Gg1} = \frac{dG}{d\tau} \frac{1}{G} \sqrt{\frac{d_{eq}}{\mu}} \tag{1.94}$$

$$K_{Gg2} = \frac{dG}{d\tau} \frac{1}{G} \sqrt{\frac{\nu}{u_f g}} \tag{1.95}$$

where ν is the kinematic viscosity coefficient.

In the majority of cases, the parameters (1.89)–(1.95), alongside with the parameters (1.69) or (1.70) and (1.81)–(1.85), will affect the Nusselt number as an unsteady change in G will be accompanied by varying T_w. The resistance coefficient ξ depends on $T_w(x, \tau)$ only in virtue of the contribution of unsteady variations of T_w to turbulence generation and a velocity profile. And since an unsteady temperature profile is greatly affected by unsteady alterations in turbulence generation and distribution over the cross-section, unsteady temperature and velocity profiles must influence each other substantially. The dimensionless parameters (1.89)–(1.95) are related in terms of the Re numbers and the parameters F_f, d_{eq}, g and v:

$$K_G = K_{G1}\text{Re}^2 = K_u = K_{u1}\text{Re} = K_{Gg1}\frac{\sqrt{gd_{eq}^3}}{v}$$

$$= K_{Gg2}\frac{d_{eq}^2}{v}\sqrt{gu_f} = K_{Gg}\frac{d_{eq}}{v}\sqrt{\frac{gu_f F_f}{v}} \quad (1.96)$$

Therefore, the functional relations for unsteady heat transfer and hydraulic resistance are supplemented, at least, with one of the parameters, (1.89)–(1.95), e.g. with K_G by formula (1.90).

Let us dwell on the effect of the variability of heat carrier properties on unsteady heat transfer. An unsteady change in a temperature profile in terms of varying thermophysical properties (ρ, c_p, λ, μ) near a wall, where turbulence is mainly generated, may affect the turbulence intensity and distribution substantially. Therefore, different dependences of the Nusselt number on varying thermophysical properties of heat carrier under steady and unsteady conditions should be expected, too.

From the aforesaid it follows that in a general case of unsteady turbulent channel flow the functional relations for Nu and ξ will be of the following form:

$$\text{Nu} = f_1\left(\frac{x}{d_{eq}}, \text{Re}_f, \text{Pr}_f, \frac{\mu_w}{\mu_f}, \frac{\lambda_w}{\lambda_f}, \frac{\rho_w}{\rho_f}, \frac{c_{pw}}{c_{pf}} K_{T\tau}, K_{Tg}^*, K_G\right) \quad (1.97)$$

$$\xi_f = f_2\left(\frac{x}{d_{eq}}, \text{Re}_f, \frac{\mu_w}{\mu_f}, \frac{\rho_w}{\rho_f}, K_{Tg}^*, K_G\right) \quad (1.98)$$

Relations (1.97) and (1.98) do not incorporate the parameters K_{Tx}, (1.71) and K_{qx}, (1.72), that slightly affect unsteady heat transfer.

It is important to emphasize that these relations do not explicitly include the time (e.g. in the form $T = \tau u_f^2/a$, Fo $= a\tau/d_{eq}^2$, or Ho $= \mu u_f \tau/d_{eq}$). The values of all the quantities in (1.97) and (1.98) are taken at one and the same time instant. The time scale does not, therefore, play an important role either in modelling or in calculating by relations (1.97) and (1.98). This situation, in which the unsteadiness may be allowed for by including only the parameters of type (1.69) and (1.89), was called by S. S. Kutateladze [22] as a local time similarity.

If the wall flux density law is prescribed, then instead of $K_{T\tau}$, (1.69), it is convenient to include, into (1.97), the parameter $K_{q\tau}$, (1.70), that is responsible for a time variation of $q_w(x, \tau)$.

If account is taken of the fact that the wall heat flux density is $q_w = \alpha(T_w - T_f)$, then

$$K_{q\tau} = \frac{\partial[\alpha(T_w - T_f)]}{\partial \tau} \frac{d_{eq}^2}{\alpha(T_w - T_f)a}$$

$$= \frac{\partial T_w}{\partial \tau} \frac{d_{eq}^2}{(T_w - T_f)a} - \frac{\partial T_f}{\partial \tau} \frac{d_{eq}^2}{(T_w - T_f)a} + \frac{\partial \alpha}{\partial \tau} \frac{d_{eq}^2}{\alpha a} = K_{T\tau} - K_{h.c} + K_\alpha \quad (1.99)$$

where

$$K_{\text{heat conduction}} = \frac{\partial T_f}{\partial \tau} \frac{d_{eq}^2}{(T_w - T_f)a}; \quad K_\alpha = \frac{\partial \alpha}{\partial \tau} \frac{d_{eq}^2}{\alpha a}$$

As seen, the parameters $K_{q\tau}$ and $K_{T\tau}$ are related.

The Nu_f, Re_f numbers entering into (1.97) as well as the values of μ, λ, c_p, ρ and ν included into the parameters $K_{T\tau}$, K_{fg}^* and K_G are determined through the mean-mass flow temperature T_f in the considered cross-section. The problem on a choice of a determining temperature is of no principal importance as relation (1.97) incorporates the parameters that allow for the flow nonisothermity.

For gases with $\text{Pr}_f = \text{const}$ the thermophysical properties mainly depend on temperature, and these relations are of the form:

$$\frac{\rho_f}{\rho_w} = \left(\frac{T_f}{T_w}\right)^{n_\rho}; \quad \frac{c_{pf}}{c_{pw}} = \left(\frac{T_f}{T_w}\right)^{n_w}; \quad \frac{\lambda_f}{\lambda_w}$$

$$= \left(\frac{T_f}{T_w}\right)^{n_\lambda}; \quad \frac{\mu_f}{\mu_w} = \left(\frac{T_f}{T_w}\right)^{n_\mu} \quad (1.100)$$

where n_ρ, n_w, n_λ and n_μ are the constants that depend on the nature of a gas and the temperature range. In the majority of cases, $n_\rho = -1$. It is, therefore, enough to include the dimensionless parameters T_w/T_f, n_w, n_λ and n_μ as well as only T_w/T_f for a specific gas into the criterial equations for gas heat transfer to allow for the contribution of the variability of thermophysical properties. As shown, for gases the impact of unsteady heat conduction taken into account by the parameter $K_{T\tau}$ or $K_{q\tau}$ is small. Functional relation (1.97) for gases, therefore, assumes the form:

$$\text{Nu}_f = f_1\left(\text{Re}_f, \frac{T_w}{T_f}, n_w, n_\lambda, n_\mu, \frac{x}{d_{eq}}, K_{fg}^*, K_G\right) \quad (1.101)$$

For a particular gas this relation is of the form

$$\mathrm{Nu}_f = f_1\left(\mathrm{Re}_f, \frac{T_w}{T_f}, \frac{x}{d_{eq}}, K_{Tg}^*, K_G\right) \qquad (1.102)$$

and the appropriate quasi-stationary relation is

$$\mathrm{Nu}_{f.qs} = f_0\left(\mathrm{Re}_f, \frac{T_w}{T_f}, \frac{x}{d_{eq}}\right) \qquad (1.103)$$

If, as shown in [24], the effect of x/d_{eq} on unsteady heat transfer is the same as the one on steady heat transfer and the contribution of Re_f and T_w/T_f to unsteady heat transfer is different, as compared to the steady one, then for gases

$$K_\alpha = \frac{\mathrm{Nu}_f}{\mathrm{Nu}_{f.qs}} = f\left(\mathrm{Re}_f, \frac{T_w}{T_f}, K_{Tg}^*, K_G\right) \qquad (1.104)$$

For drop liquids, with varying a temperature the viscosity changes most strongly, as compared to other properties, therefore, the effect of the variability of properties on heat transfer is usually taken into consideration by the ratio μ_f/μ_w or $\mathrm{Pr}_f/\mathrm{Pr}_w$. In this case, for drop liquids relation (1.97) assumes the form:

$$\mathrm{Nu}_f = f_1\left(\mathrm{Re}_f, \mathrm{Pr}_f, \frac{\mathrm{Pr}_f}{\mathrm{Pr}_w}, \frac{x}{d_{eq}}, K_{T\tau}, K_{Tg}^*, K_G\right) \qquad (1.105)$$

and the appropriate quasi-stationary relation is

$$\mathrm{Nu}_{f.qs} = f_0\left(\mathrm{Re}_f, \mathrm{Pr}_f, \frac{\mathrm{Pr}_f}{\mathrm{Pr}_w}, \frac{x}{d_{eq}}\right) \qquad (1.106)$$

If, as shown in [24], the effect of x/d_{eq} on unsteady heat transfer of liquids is the same as on steady heat transfer, then it may be obtained:

$$K_\alpha = \frac{\mathrm{Nu}_f}{\mathrm{Nu}_{f.qs}} = f\left(\mathrm{Re}_f, \mathrm{Pr}_f, \frac{\mathrm{Pr}_f}{\mathrm{Pr}_w}, K_{T\tau}, K_{Tg}^*, K_G\right) \qquad (1.107)$$

Relation (1.107) may be represented as

$$K_\alpha = K_{\alpha 1}(K_{T\tau}, \mathrm{Re}_f, \mathrm{Pr}_f)\, K_{\alpha 2}(K_{Tg}^*, \mathrm{Re}_f)\, K_{\alpha 3}(K_G, \mathrm{Re}_f) \qquad (1.108)$$

or, as shown in [24], as

$$K_\alpha = 1 + \Delta K_{\alpha 1}(K_{T\tau}, \mathrm{Re}_f, \mathrm{Pr}_f) + \Delta K_{\alpha 2}(K_{Tg}^*, \mathrm{Re}_f)$$
$$+ \Delta K_{\alpha 3}(K_G, \mathrm{Re}_f) \qquad (1.109)$$

where $K_{\alpha 1}$, $K_{\alpha 2}$ and $K_{\alpha 3}$ are the unsteady heat transfer-to-quasi-stationary coefficient ratios due to unsteady heat conduction superimposition on steady convective heat transfer, due to varying turbulent flow structure with increasing or decreasing T_w and due to flow acceleration or deceleration, respectively; $\Delta K_{\alpha 1}$, $\Delta K_{\alpha 2}$ and $\Delta K_{\alpha 3}$ are the corresponding variations of K_α. As mentioned in [24], the ratio ($\mathrm{Pr_f/Pr_w}$) does not exert influence on K_α.

For gases $K_{\alpha 1} = 1$ and $\Delta K_{\alpha 1} = 0$, and relation (1.97) assumes the form

$$K_\alpha = K_{\alpha 2}\left(K_{\mathrm{f}g}^*, \mathrm{Re_f}, \frac{T_w}{T_f}\right) \cdot K_{\alpha 3}\left(K_G, \mathrm{Re_f}, \frac{T_w}{T_f}\right) \quad (1.110)$$

or similarly (1.109)

$$K_\alpha = 1 + \Delta K_{\alpha 2}\left(K_{\mathrm{f}g}^*, \mathrm{Re_f}, \frac{T_w}{T_f}\right) + \Delta K_{\alpha 3}\left(K_G, \mathrm{Re_f}, \frac{T_w}{T_f}\right) \quad (1.111)$$

If from experiment (or in solving three-dimensional problems) empirical relations (1.102), (1.105) or (1.104), (1.107) are found, then use of a one-dimensional approach to making engineering calculations of unsteady thermal processes will be similarly effective as in the case of steady processes. In this case, solving unsteady heat conduction problem (1.63) by the successive approximation methods under the third-kind boundary conditions

$$T_w = T_f + \frac{q_w}{\alpha} = T_f - \frac{\lambda_w \dfrac{\partial T}{\partial n}}{\alpha} \quad (1.112)$$

where n is the normal to the wall surface at a given point, offers no difficulties.

1.4 SPECIFIC FEATURES OF UNSTEADY HEAT AND MASS TRANSFER PROCESSES

In Section 1.3 it has been emphasized that the variability of thermophysical properties of liquid and the unsteadiness effect on flow turbulence should be allowed for to calculate unsteady heat transfer. In this case, the heat transfer coefficient as a function of unsteady boundary conditions is determined from experiment.

Similar problems are also being solved in studying unsteady heat and mass transfer in helical tube bundles. Here consideration is made of the processes of forming and re-arranging heat carrier temperature fields in time and in a solid phase (of helical tubes) at nonuniform heat supply in the bundle cross-section. However, if in investigating heat transfer in helical tube bundles uniformly heated in the bundle cross-section, the conjugation problem can be solved using a one-dimensional description of the processes, then in analyzing unsteady temperature fields at nonuniform heat supply in the bundle cross-section, it is necessary to solve either an axisymmetric or a three-dimensional problem within the framework of the homogeneized flow model (Section 1.2). In considering the axisymmetric problem, the system of equations (1.36)–(1.40) is being solved, in a general case, under boundary conditions (1.41)–(1.44). This is attributed to the fact that the system of the equations of gasdynamics with respect to the x-coordinate is hyperbolic. Therefore, when air is flowing with a subsonic velocity, the disturbances due to the process unsteadiness propagate both in the flow and backflow directions, thereby arising a necessity to prescribe time variations of total ($p_{t,in}$) and static (p_{out}) pressures, (1.41) and (1.42), respectively. The problem may be simplified if the time of establishing a quasi-stationary gasdynamic process is compared with the one, for which weak disturbances pass through a helical tube bundle. Disturbance waves in a gas are propagating with a sound velocity, and the time of establishing a quasi-stationary gasdynamic regime in the considered helical tubes is approximately 0.1 s. Therefore, in a number of cases, instead of equations (1.38) and (1.39), it is possible to use the stationary equations of gasdynamics with simplified boundary conditions by assigning a time variation of heat carrier flow rate $G = G(\tau)$ and by considering it constant along the channel at each computational time moment (Chapter 5). This applies to the so-called hydrodynamic unsteadiness of a process when a heat carrier flow rate varies in time and supplied heat power remains constant. In the case of the thermal unsteadiness of a process, when a heat carrier flow rate through a bundle remains constant in time and heat power varies (start of a bundle, transition from one operating regime to another, stop of an apparatus), the stationary equations of gasdynamics may be also used instead of equations (1.38) and (1.39).

In developing a numerical method to solve the system of equations (1.36)–(1.40), it was necessary to allow for the specific features associated with solving the gasdynamics problems numerically. Thus, the gasdynamics equations are nonlinear, and the theory of the difference methods has been mainly developed for linear problems. Therefore, this system was beforehand quasi-linearized, i.e. the coefficients at the derivatives were averaged depending on the differentiation coordinate and were removed from the differentiation sign. These coefficients were refined in the iteration cycles. Since the methods to solve heat transfer equations (1.36) and (1.37) and the system of motion (1.38) and continuity (1.39) equations were different, the problem was solved at two successive stages: first, heat transfer equations were solved (thermal part of the problem) and then motion and continuity equations were solved simultaneously (gasdynamic part of the problem). The solutions of these parts of the problem were related in terms of the state equation and iteration cycles. As an explicit scheme was adopted to solve equations (1.38) and (1.39), this imposed the stringent restrictions on a choice of the time step $\Delta\tau/\Delta x \leq 1/(u + a)$. A great deal of computer time is needed to solve such a problem. Therefore, using the operating conditions of a helical tube bundle, when a time variation of gas velocities, densities and pressures is much greater than the transit time of weak disturbances, the gasdynamic parameters at any time instant τ may be calculated by the stationary methods. In this case, the time step is $\Delta\tau \gg 10^{-3}-10^{-4}$s. The variable-direction method involving an implicit scheme being stable over a wide range of space-time steps was utilized to solve the thermal part of the problem and does not require stringent restrictions on a choice of a time step.

The homogeneized flow model for unsteady heat and mass transfer in a helical tube bundle (Section 1.2), its mathematical description and the specific features of the method of solving the problem are substantiated from experiment by comparing the predicted and experimental temperature fields of heat carrier in a real helical tube bundle. In this case, the validity of the simplifying assumptions made at a mathematical description of the problem is supported, and the possibility is shown to close the system of equations (1.36)–(1.40) using the effective diffusion coefficient K_{uns}. In experimental studies of the coefficient K_{uns}, account is taken of the fact, how K_{uns} is affected by all the transfer mechanisms characteristic of helical tube bundle flow under steady and unsteady conditions as well as the limits of use of the quasi-stationary value of this coefficient are determined when applied to calculate unsteady temperature fields of heat carrier.

The specific features of heat and mass transfer in a helical tube bundle are specified both by the chief characteristic of its design and by the unsteadiness impact. As mentioned, the flow swirling in a helical tube bundle is determined by the relative pitch S/d or by the number

$$\mathrm{Fr}_M = \frac{S^2}{dd_{eq}} \tag{1.113}$$

characterizing a relationship between the inertia and centrifugal forces that act upon the flow subject to swirling. The Fr_M number represents a complex geometrical characteristic of a bundle. The less is S/d (or Fr_M), the greater is the flow swirling intensity. The flow swirling manifests itself, first of all, in the fact that in the bundle cross-section there appear fields of transverse velocity vector components v and w which are plotted in Fig. 1.6a, b. With decreasing Fr_M, the velocity v parallel to the wider side of the oval tube profile increases (Fig. 1.6a), the direction of this velocity being determined through that of the helical tube twisting. At the same direction of the helical tube twisting the velocity at the tube boundaries is v = 0, which is specified by the laws for interacting vortices.

In the wall layer of the tube the velocity v varies following the quasi-solid rotation law [39], a maximum value of the velocity v being set at the external boundary of the wall layer. Thus, the velocity v varies, in a thin wall layer, from zero on the tube wall to its maximum at the external boundary. With increasing the Reynolds number at a prescribed Fr_M number, the

34 UNSTEADY HEAT AND MASS TRANSFER IN HELICAL TUBE BUNDLES

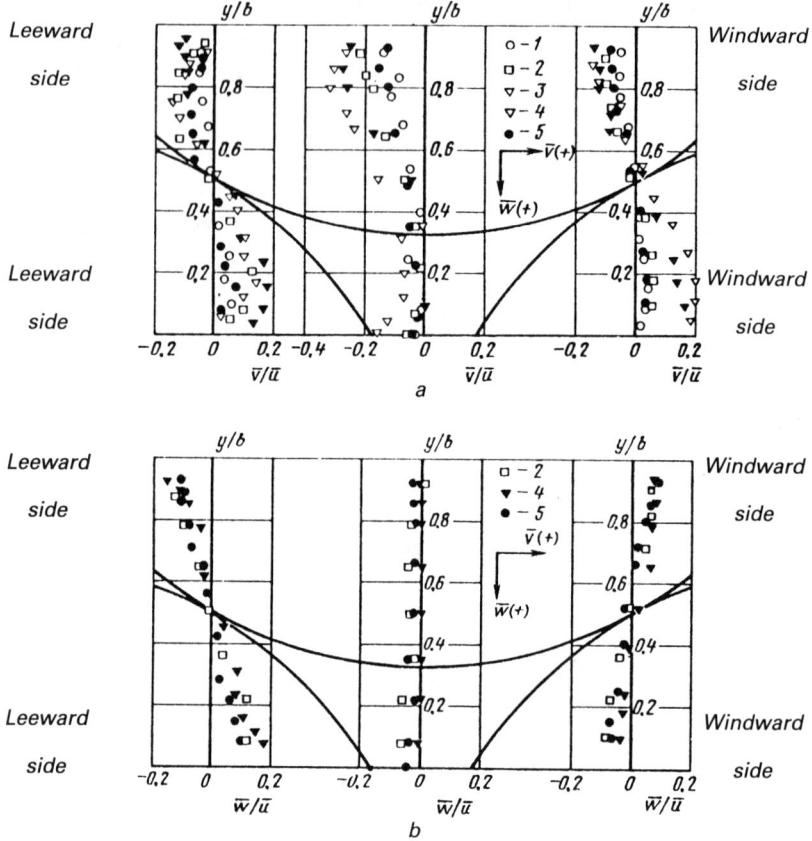

Figure 1.6 Distributions of the transverse velocity components v (*a*) and w (*b*) in a bundle cell: ○, □, ▽, experimental data for bundles with Fr_M = 1187, 296 and 178 and with Re = $1.5 \cdot 10^4$; ▼, ●, the same for a bundle with Fr_M = 296 and with Re = $5.8 \cdot 10^3$; $6.7 \cdot 10^4$.

flow swirling intensity decreases and, hence, the velocity v decreases, too (Fig. 1.6*b*). Therefore, in the transition range of the Re numbers $\leqslant 10^4$ the greater intensity of heat and mass transfer processes should be expected. The velocity vector component w normal to the wider side of the oval tube profile as well as the velocity component v attains its maximum value at the external boundary of the wall layer (Fig. 1.6*b*). In this case, the velocity w in the leeward part of the profile is directed to the tube wall and in the windward one, from the wall. Such velocity curves in the cells of a helical tube bundle point to the available transfer processes between the wall layer and the flow core due to convection. A change in the velocities v and w in a thin layer from 0 to their maximum values means that the flow swirling, first of all, affects the wall region of flow where it substantially increases the turbulence level, as compared to the one in the flow core of a bundle [39]. This impact exerts an influence on an increase of the heat transfer coefficient in helical tube bundles which grows in the same proportion as the hydraulic resistance coefficient at $Fr_M \geqslant 90$ and $Re \gg 7 \cdot 10^3$ [39]. In the transition range of $Re \ll 7 \cdot 10^3$ the leading growth of heat transfer coefficients is observed, as compared to the one of the hydraulic resistance coefficient [52].

It should be noted that according to the data [25] dealing with the generalization of the results of numerous authors, for the developed turbulent flow in a smooth tube the turbulence level or Karman's number $K = \sqrt{v'^2}/u$ is, on the average, equal to 0.067 with a scatter of the points 0.05 to 0.08 and, on the average, amounts to 0.04 on the flow axis irrespective of the Reynolds number $Re = 8 \cdot 10^3 - 10^8$. These values of the Karman number have proved to be equal to those of the Euler number for an amplitude of a dynamic pressure pulsation in the appropriate regions of the tube flow

$$E' = \frac{2H'}{\rho u^2}$$

where H' is the range or double amplitude of a dynamic pressure pulsation and $\rho u^2/2$ is the velocity head. A decrease in pulsational characteristics of the flow from the tube wall to its axis have been also established by other investigators. In a helical tube bundle with $Fr_M = 178$ and $Re = 3.6 \cdot 10^4$, in [39], the turbulence level or the Karman number has proved to be over 3.3–4.3 times as large as that in a tube [25], and in the flow core of a bundle, over 1.87 times as large as that on the tube axis [25]. Thus, in a circular channel the turbulence level in a wall layer is over 1.67 times as large as that on the tube axis [25] and in a helical tube bundle it is over 2.94–3.86 times in a wall layer as large as that in the flow core. This means that in a helical tube bundle due to flow swirling the wall layer is turbulized to a greater extent than the flow core, the turbulization ratio in these flow regions at $Fr_M = 178$ is over approximately 2 times as large as the similar ratio for a round tube.

This specific feature of heat and mass transfer in a helical tube bundle associated with a contribution of flow swirling to turbulence in a wall layer is, apparently, predominant in improvement of unsteady heat transfer from the wall layer to the flow core. Turbulence generation in the wall layer due to flow swirling and intensive convective transfer of liquid masses between the wall layer and the flow core may potentiate the effect of unsteadiness on the flow structure.

A process of levelling unsteady temperature fields due to nonuniform heat supply over the radius is also attributed to the action of such a mechanism as convective ordered liquid transfer via the spiral channels relative to the tube axis. The intensity of this mechanism is specified by the relative helical tube twisting pitch S/d or by the Fr_M number. The smaller is the Fr_M number, the more intensive is the levelling of temperature field nonuniformities of the heat carrier in a bundle cross-section. This transfer mechanism is the more effective, the higher is the turbulence level at the boundary of the adjacent spiral channels in a cell. The flow core turbulence is initiated not only due to diffusion of the turbulence generation region near the tube walls but also due to tangential velocity discontinuity at the boundaries of the spiral and through channels where the flow is moving either with different velocities or has different direction [39]. Moreover, in the case of flow past the places of the contacting tubes, there appears turbulence characteristic of a wake which is formed in flow past a body. Heat and mass transfer between the bundle cells also takes place due to convection in the bundle cross-section (Fig. 1.6). These mechanisms for heat and mass transfer in a helical tube bundle promote the levelling of temperature field nonuniformities in the bundle cross-section both under steady and unsteady heat and mass transfer conditions at nonuniform heat supply (heat release) over the radius and azimuth of a bundle. This process is taken into account, in the energy and motion equations, by the diffusional terms incorporating the effective diffusion coefficient D_t determined from experiment. Criterial relations are established experimentally to use the transfer coefficient D_t or K for closing the systems of equations describing unsteady and steady flows in a helical tube bundle in the homogenized flow statement. Besides the Fr_M number, (1.113), the main numbers include the Reynolds number

$$Re = \frac{u_{mean} d_{eq} \rho}{\mu} \qquad (1.114)$$

as well as the bundle porosity with respect to heat carrier characterizing a fraction of the area of radial sections, on which there occurs heat carrier contact at the boundaries of the adjacent cells of a bundle [39]:

$$m = \frac{F_f}{F_\Sigma} \qquad (1.115)$$

where $F_\Sigma = F_f + F$ is the cross-sectional area of a bundle. Then the criterial relation for a stabilized value of the coefficient K_{stab} under steady conditions assumes the form [39]

$$K_{stab} = K(Re, Fr_M, m) \qquad (1.116)$$

Relation (1.116) is found from experiment. A choice of the equivalent bundle diameter

$$d_{eq} = \frac{4 F_f}{\Pi_{sm}} \qquad (1.117)$$

as the main linear size is attributed to the fact that in a helical tube bundle the thickness of the wall layer is small, as compared to the sizes of bundle cells and to the tube curvature radius, and inconsiderably varies over their perimeter. The possibility to use d_{eq}, in this case has been also supported by the experimental data on heat transfer and flow [39]. Together with d_{eq}, in studying heat transfer and hydraulic resistance, as the characteristic size it is possible to use the effective wall layer thickness δ taken constant over the perimeter of helical tubes and the same for thermal and hydrodynamic wall layers, which is valid for Pr ≈ 1 [10]. Use of d_{eq} in a helical tube bundle is also justified by the fact that the stagnation zones with laminar flow are not developed even at the points of contacting adjacent tubes, which is observed in close-packed bundles of round tubes [4]. Studies of the flow structure in helical tube bundles have evidenced that turbulent flow with a high turbulence level occurs at the places of contacting tubes in a wake. At the same time the area of the points of contacting the tubes is less than 3% of the total tube surface.

The Fr_M number, (1.113), has been derived, assuming that in spiral bundle channels the heat carrier flow is swirled according to the quasi-solid rotation law $v_\tau r^{-1}$ = const and that the field of centrifugal forces is estimated by the acceleration found in terms of a maximum tangetial velocity of flow swirling. In this case, the Fr_M number may be related to the number which is, in form, similar to the Froude number where the acceleration of gravity is replaced by the one of centrifugal forces [39]. The Fr_M number allows for the specific features of flow and heat and mass transfer in helical tube bundles. The porosity m mainly affects interchannel mixing alone. The smaller is m, the less is heat and mass transfer in the transverse direction.

The characteristic properties of heat and mass transfer in a twisted helical tube bundle under steady conditions are attributed to the adopted laws for the twisting of the helical tubes relative to the bundle axis. So, when the helical tubes are twisted over a bundle radius with a constant twisting angle γ = const (r), the flow is additionally swirled according to the law v_τ = const, i.e. the tangetial flow velocity on different radii of a bundle will be the same. In this case, irrespective of the bundle radius, it may be expected that the twisting of helical tubes relative to the bundle axis will have no impact on the coefficient K if azimuthal transfer is taken into account by the convective terms in the motion and energy equations (Section 1.2).

In case, the helical tubes are twisted relative to the bundle axis according to the law S_{tw} = const (r), when the heat carrier flow is swirled additionally according the quasi-solid rotation law, then the coefficient K must be a function of bundle radius since following such a law for flow swirling the turbulence level sharply grows in the axial region of a bundle and falls off in the peripheral zone of a bundle. A similar situation is observed near the axial-blade swirlers [47].

The specific features of unsteady heat and mass transfer are defined not only by flow swirling but also attributed to varying the turbulent flow structure in the wall flow region. The unsteady heat and mass transfer mechanism in this region will be mainly specified by the same process as in the case of unsteady heat transfer in round tubes [24]. This mechanism has been examined in Section 1.3. As mentioned, unsteady temperature field of heat carrier in a helical tube bundle are affected by the transfer mechanisms typical of steady mixing processes of heat carrier.

If the analysis is based on the methods of the similarity theory, then from a complete system of equations describing unsteady heat transfer from the liquid flow it is possible to determine the Fourier number (heat time homogeneity) characterizing a relationship between a rate of varying a temperature field of heat carrier, its physical properties and sizes of the flow region [27]:

$$Fo_b = \frac{\lambda_B \tau}{c_p \rho_b d_{sh}^2} \tag{1.118}$$

and being the main number used to describe unsteady processes with the aid of the criterial relations. Then the dimensionless effective unsteady diffusion coefficient will be determined by the following criterial relation:

$$K_{uns} = K(Re, Fr_M, m, Fo_b) \tag{1.119}$$

This relation must be found from experiment. The concept of the relative mixing coefficient $\kappa = K_{uns}/K_{qs}$ where K_{qs} is the quasi-stationary value of the dimensionless effective turbulent diffusion coefficient is introduced, giving a more convenient form to present the experimental data on unsteady heat and mass transfer. Then the criterial relation will be of the form:

$$\kappa = \kappa(Fo_b, Fr_M, Re) \tag{1.120}$$

In this case, (1.120) does not incorporate the bundle porosity m with respect to the heat carrier as this quantity equally produces an effect on the coefficients K_{uns} and K_{qs} for the considered helical tube bundle. Criterial equation (1.120) may also include the parameters responsible for the effect of unsteady boundary conditions on the coefficient κ. These are the unsteadiness criteria, parameters of the type $\partial N/\partial \tau$, etc. The impact of different factors on unsteady heat and mass transfer in helical tube bundles will be detailed in Chapter 5.

In experimental studies of unsteady and steady heat and mass transfer processes much attention has been paid to extend the modelling potentialities and to apply the experimental data obtained under the same conditions for designing specific heat exchangers and devices. The present book uses the results on the flow structure to explain and to analyze the transfer mechanisms in helical tube bundles and new revealed effects as well as it considers different data processing and generalization methods.

Different methods allowing for the specific features of the designs of in-line flow helical tube bundles and those of unsteady heat and mass transfer processes have been worked out to make experimental studies.

To investigate unsteady heat and mass transfer, it is necessary to reveal the effect of different factors on transfer coefficients to analyze the flow structure both under steady and unsteady conditions, to employ the newest measuring and experiment control facilities as well as computer data acquisition and processing devices. Use of such experiment automatic control systems enables one to improve the reliability of the obtained results.

Experimental and theoretical methods of solving this problem supplement each other and permit one to make reliable recommendations on calculations of unsteady and steady temperature fields.

CHAPTER
TWO
METHODS OF EXPERIMENTAL STUDY OF HEAT CARRIER MIXING

2.1 METHODS OF EXPERIMENTAL STUDY OF HEAT CARRIER MIXING UNDER UNSTEADY AND STEADY CONDITIONS

The methods of experimental study of heat carrier mixing in the helical tube bundle cross-section under steady conditions were analyzed in [39]. These are the classical methods of studying transport properties of flow: methods of heat (substance) diffusion from a point source continuously emitting heated air (or other gas) particles into the main flow and from a linear source which are transformed, considering the specific properties of flow in a helical tube bundle as well as of a bundle design. In this case, the homogenized flow model was adopted to make experiments and to process the data. Temperature fields and flow velocities were measured outside the wall layer, and the predicted temperature fields of heat carrier and flow velocities were continuous within the bundle shell diameter. In addition, it was assumed that a two-phase homogenized medium with a fixed solid phase was flowing in a bundle. In studying the effective turbulent diffusion coefficient in a straight helical tube bundle by the first method, the diffusion source diameter was equal to the helical tube one d, and the source travelled relative to the outlet cross-section of a bundle where velocity and temperature fields were measured. However, these deviations from the known diffusion method did not hinder use of the concept of a point source in a helical tube bundle at rather large distances from it where the measured temperature fields practically did not differ from the Gauss distribution [39]. This method based on the statistical Lagrange representation of a turbulent field when applied to examine a history of the motion of individual particles continuously emitted by a diffusion source is also employed in the present monograph to determine effective turbulent diffusion coefficients in a twisted bundle of helical tubes but in the presence of fixed diffusion sources.

The method of heat diffusion from a fixed diffusion source in the field of equal velocities is based on using the limiting solutions to Taylor's equation for homogeneous and isotropic turbulence ($\sqrt{\overline{u'^2}} = \sqrt{\overline{v'^2}} = \sqrt{\overline{w'^2}} = v_1$)

$$\frac{d}{dt}\left(\frac{1}{2}\overline{y}^2\right) = \int_{t_0}^{t} \overline{v_1(t_0)v_1(t')}\, dt' \tag{2.1}$$

It is known that in the turbulent flow lacking any regular pattern, two arbitrary particles are moving so as to increase a distance between them in due time. In the presence of a great number of particles emitted by a diffusion source, in some time one particle is travelling in one direction and the other, in the opposite direction so that in the case of isotropic and homogeneous turbulence the particle propagation will be symmetric relative to the source axis. The same situation has been observed in experiments, in which temperature fields have been measured in the cross-section of a bundle [39]. Time turbulence homogeneity, i.e. a turbulent field with the same values of $\sqrt{\overline{v'^2}}$ at all the points, enables one, in this case, to make averaging over a great number of particles which start moving at successive time instants. A distance covered by a labelled particle at cross mixing of heat carrier in the y-direction from the initial moment $t_0 = 0$ to the moment t is equal to

$$y(t) = \int_{0}^{t} v_1(t')\, dt' \tag{2.2}$$

and an expression for the mean-statistical squared displacement is determined as:

$$\frac{1}{2}\frac{d}{dt}\overline{y}^2 = \overline{y\frac{dy}{dt}} = \overline{yv_1(t)} = \overline{v_1(t)\int_{0}^{t} v_1(t')\, dt'} \tag{2.3}$$

A bar in expression (2.3) stands for averaging. If according to Taylor's theory the time is divided into n intervals and summation is made over all the particles at each time interval, then a mean value of the product is equal to

$$\overline{v_1(t)\int_{0}^{t} v_1(t')\, dt'} = v_1^2\, R_\tau \tag{2.4}$$

where R_τ is the coefficient for time correlation of velocities at the moments t and $(p/n)\,t$ where p is the ordinal number of some time interval since the quantity v_1^2 has one and the same value at the moments t and $(p/n)\,t$. If a time interval between these moments is designated through τ, and pass is made to the limit with increasing n, then the element $d\tau$ may be substituted for t/n, and a mean value of the product in (2.4) will be equal to $v_1^2 R_\tau \, d\tau$. Then

$$\frac{d}{dt}\left(\frac{1}{2}\overline{y^2}\right) = v_1^2 \int_0^t R_\tau \, d\tau \tag{2.5}$$

At a large diffusion time the coefficient $D_t = v_1 L = v_1^2 = v_1^2 \int_0^t R_\tau d\tau$ where L is the spatial scale of eddy diffusion, and upon integration of expression (2.5) we have

$$\overline{y^2} = 2D_t t + \text{const} = \frac{2D_t x}{u} + \text{const} \tag{2.6}$$

In expression (2.6), an assumption is made that the diffusion time $t = x/u$. This condition is valid if pulsation velocities are not large as against the flow velocity u. This condition is satisfied, to a first approximation, for helical tube bundles. On the y^2- and x-coordinates, straight line (2.6) gives the intercept x_0 on the abscissa axis, and the constant in equation (2.6) is equal to $2D_t x_0/u$. Then, from experiment, knowing a heat carrier temperature distribution at different distances from a diffusion source and having found the quantity $\bar{y}^2$ for each distribution, the coefficient D_t may be obtained from the limiting solution to Taylor's equation (2.1) at a large diffusion time

$$\overline{y^2} = 2\frac{D_t}{u}(x - x_0) \tag{2.7}$$

Experiments [39] have revealed that dimensionless excess temperature distributions in a helical tube bundle obey the Gauss distribution

$$\frac{T - T_0}{T_{\max} - T_0} = \exp\left(-\frac{y^2}{2\bar{y}^2}\right) \tag{2.8}$$

where T_0 is the main flow temperature and $T_{\max}$ is the maximum axial temperature of the jet flowing from the diffusion source. This effect displays a general rule as at large distances from the diffusion source the Gauss distribution always succeeds statistically in setting up irrespective of pulsation velocity distributions. At small distances from the diffusion source (at a small diffusion time) the limiting solution to Taylor's equation (2.1) assumes the form

$$\overline{y^2} = v_1^2 t = \frac{v_1^2}{u^2} x^2 \tag{2.9}$$

However, this relationship is not practically implemented in helical tube bundles [39]. This may be attributed to the impact of finite sizes of the diffusion source, velocity field nonuniformity in the flow core and to the blocking of the flow under study with helical tubes. This promotes the heated particles near the jet outlet to cover a great number of different non-correlated distances from the diffusion source to the considered point although velocity pulsation distributions in the flow core at $Re > 10^4$ also approach a normal distribution law. At $Re < 10^4$ the velocity pulsations deviate from the Gauss law in a helical tube bundle, pointing to turbulence anisotropy in such bundles over this range of Re numbers. Therefore, in a twisted bundle of helical tubes the method of heat diffusion from a source was used only to determine the coefficient D_t, and its use was confirmed by agreement between the experimental temperature distributions and the Gauss distribution although the basic assumptions of the Taylor theory in this case are not violated rigorously. In experiments, the radius of a diffusion source exceeds approximately 3 times the helical tube one. In this case, the properties of the tracer gas (heated air) and of the main flow are the same. This enables one to obtain rather reliable experimental data on the coefficient D_t. At the same time if in [39] for a straight bundle of helical tubes where the source radius was equal to the helical tube one we have succeeded in estimating the turbulence intensity by equation (2.9), then in this situation this cannot be done because of large sizes of a source. To improve the accuracy in determining the coefficient D_t, experiments on heat carrier mixing in a twisted bundle were made at a fixed diffusion source. The thermocouples were embedded in the helical tubes to determine temperature fields at different distances from a source. In addition, measurements were made of a tube wall temperature (i.e. a solid phase temperature in terms of the homogenized flow model). This measuring procedure could promote errors in determining the coefficient D_t since the temperature distributions in the heat carrier flow core and on the tube wall are different. Hence, the mean-statistical squared displacement $\bar{y}^2$ and D_t are different, too. This difference is, apparently, systematic in nature. Section 4.2 deals with the approach, considering a correction for the determined coefficient D_t in wall temperature measurements.

When the method of diffusion from a point source was adopted to determine the coefficient D_t in a twisted bundle of helical tubes, it was necessary also to allow for a distinction bound up with the curvilinearity of the axis of a heated jet injected into a bundle where a temperature in each bundle cross-section was maximum. In addition, the swirling angle of the jet axis was equal to the twisting one of helical tubes on the appropriate bundle radius. Therefore, the experimental coefficient D_t in this case does not take into account the effect of flow swirling by a twisted bundle in the azimuthal direction, and it may be compared with the coefficient D_t for a straight helical tube bundle. This enables one to reveal the effects, peculiar to flow in a twisted bundle, on heat and mass transfer in this bundle. At the same time the impact of flow swirling by a twisted bundle on heat and mass transfer in the azimuthal direction can be allowed for by including, into the motion and energy equations used for calculating temperature and velocity fields in a bundle, the terms responsible for azimuthal convective ordered transfer due to twisting of a helical tube bundle relative to the bundle axis according to the prescribed law.

The method of diffusion from a system of linear heat sources for the first time employed to study steady mixing in such bundles [9] was applied to investigate unsteady mixing of heat carrier in a helical tube bundle. This method consists in examining heat diffusion from a group of heated tubes downstream. Usually the groups of 7 and 37 helical tubes [39] were heated depending on a total number of tubes in an experimental bundle.

The central zone composed of 37 helical tubes was heated to study unsteady heat and mass transfer on the bundles consisting of 127 tubes. The tubes with a high Ohmic resistance were heated by electric current. The developed nonuniformity of heat release over the bundle radius initiates the one of the temperature fields of heat carrier. Air served as heat carrier. The temperature nonuniformity was partly levelled due to cross interchannel mixing of heat carrier. This process showed the effective diffusion coefficient D_t which was found from comparison of the experimental and predicted temperature fields in the framework of the accepted model for a homogeneized medium which took the place of the heat carrier flow in a real tube bundle.

Electric insulation of the central zone of the helical tubes from the non-heated ones of a bundle was provided for ease of implementation of the method of heat diffusion in a bundle, and the velocity and temperature fields of heat carrier were measured in the flow core (outside the wall layer).

Use of the method of heat diffusion from a system of linear heat sources when applied to determine the coefficient D_t under unsteady conditions has its specific features. This is, first of all, bound up with a necessity to consider, in a general case, the problem in the conjugated statement as the heat transfer processes in heat carrier and in tube walls are interrelated, and the conditions at the boundary with heat carrier are unknown. When the homogenized flow model is used, one succeeds not to determine temperature fields in the tube walls and to prescribe beforehand boundary conditions using the concept of the boundary condition-dependent heat transfer coefficient. In this case, the thermal inertia of helical tubes is allowed for by including the heat conduction equation for a solid phase into a system of the equations describing unsteady heat and mass transfer in a bundle, and time and space temperature variations of tubes are identical to those of the solid phase of a homogenized medium. The system of equations (1.36)–(1.40) inspected in Chapter 1 enables one to calculate temperature fields of heat carrier and tube wall (solid phase) which depend on the longitudinal and radial coordinates at different time instants, i.e. to solve a two-dimensional unsteady problem. Chapter 5 will be concerned with a system of equations and a method of its solution which also allow solving the problem on asymmetric nonuniform heat supply. However, as the analyses of steady three-dimensional and axisymmetric problems have evidenced, the coefficient D_t found for these flows remains invariable under other conditions being equal. It is, therefore, advisable that in studying unsteady heat and mass transfer in helical tube bundles, consideration be restricted to the axisymmetric problem alone. Such a problem has been solved for the first time since all the previous investigations have been confined by using a one-dimensional description of unsteady heat transfer processes in channels when analysis is made of flow with a channel cross-section-constant velocity and temperature which vary along the channel alone. In this case, a wall temperature is found from the Newton equation for a heat flux in terms of the experimental values of the heat transfer coefficient [24, 26].

The distinctive feature of use of the method of heat diffusion from linear heat sources when applied to study unsteady heat carrier mixing is also bound up with a necessity to measure heat power and temperature fields of heat carrier varying in time with a large velocity. As a result, there has arisen a necessity to design a quick-response automatic-control system for starting a heat exchanger and a recording system for data acquisition and processing. Only in this case, it has become possible to determine experimentally rather reliable values of the coefficient D_t.

Non-stationary values of the coefficient D_t are also found by comparing the measured and predicted temperature fields of heat carrier for each time instant at different preliminary prescribed values of the coefficient D_t, provided that a scatter of the predicted temperature fields on the curves $T = T(r, K)$ incorporates the experimentally measured values of heat carrier temperatures. In this case, quite a certain value of the coefficient $K = D_t/(ud_{eq})$ may be assigned to each experimental point on the curve $T = T(r, K)$ according to a system of the plotted curves $T = T(r, K)$. Then for each design curve at a given value of K it is possible to find a square root of a sum of the squared deviations of each experimental point from this curve $\sqrt{\sum_{i=1}^{n} (\delta T_i)^2}$ and to plot the relation

$$\sqrt{\sum_{i=1}^{n} (\delta T_i)^2} = f(K) \qquad (2.10)$$

where n is the number of experimental points. A minimum of function (2.10) corresponds to a maximum confidence value of the dimensionless coefficient K, at which the best coincidence between the experimental and predicted temperature fields is attained.

When the non-stationary values of the coefficient D_t are determined by this method, a measuring accuracy of temperatures may be different. So, with sharply increasing heat power, a scatter of the theoretical curves $T = T(r, K)$ at different K may be small at the first time moments and can be compared with the measuring error of a temperature. Therefore, that the coefficient K_{uns} be determined with a permissible accuracy of 25–30%, account should be taken of the experimental data on those time moments, at which a scatter of the predicted temperature fields is substantially greater than the measuring error of the heat carrier temperature.

In studies of unsteady mixing the combs of thermocouples must be used to measure temperature fields over a bundle radius. In this case, the thermocouples indicate a time variation of a temperature at each specific point of flow around a bundle. These points are chosen in the flow core, the majority of the points being located in the heated bundle zone specified by the highest scatter of the predicted temperature fields. It is necessary, too that the flow parameters under unsteady conditions be measured by low-inertia transducers. So, the thermocouples must be made of a small-diameter wire, i.e. these possess small thermal inertia, to fix, with a sufficient accuracy, a real temperature of heat carrier at each time moment.

Special experimental set-ups and sections with a computerized automatic control, data acquisition and data processing system have been worked out and designed to realize the above methods of experimental study of heat carrier mixing.

2.2 EXPERIMENTAL SET-UPS FOR STUDYING HEAT CARRIER MIXING

Unsteady mixing of heat carrier was studied experimentally on the same set-up as in the case of a steady process. A central group of a 37-helical tube bundle was heated, and the helical tubes were, in turn, electrically insulated from the non-heated ones with glass-fibrous cloth that covered the tubes in the form of a fire-proof silicate-organic varnish-coated shell. A schematic of this set-up is shown in Fig. 2.1. It represents an aerodynamic open-type loop. Air is supplied to the loop by means of the turbo-compressor at an output up to 3600 m³/h (up to 1 kg/s) and is cooled in the cooler. An auxiliary line of the compressor station may be connected with the output

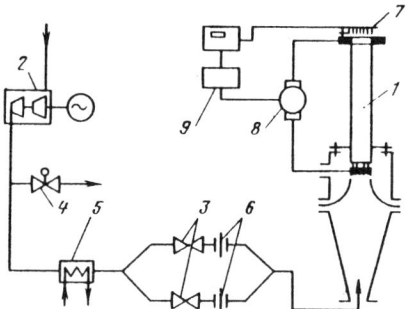

Figure 2.1 Schematic of the experimental set-up: *1*) experimental section; *2*) turbo-compressor; *3*) shut-off cock; *4*) throttle cock; *5*) cooler; *6*) orifice plates; *7*) traverse gear provided with thermocouples; *8*) generator; *9*) automatic control and measuring system.

line of the turbo-compressor to provide an air flow rate up to 1.4 kg/s. In the cooler the turbo-compressor air is cooled up to 40–60° C and then enters the experimental section via one of two parallel mains equipped with orifice plates. In one main, double diaphragms are set to provide air flow rate measurements over the range of 300–1500 kg/h and in another, over the range of 900–4300 kg/h. Diaphragms are manufactured and set according to the requirements for flow rate measurements using this practice. A diaphragm pressure drop is measured by a water well-type manometer and a pressure in front of the diaphragms, by the standard one. Air flow rate is regulated by the remote-acting gate valves mounted behind the turbo-compressor in the basic and by-pass mains as well as in those equipped with the orifice plates, providing a smooth flow rate. Air enters the experimental sections from bottom upwards via a diffusor, a system of levelling grids and Vitoshinsky's nozzle. A diameter of the narrow part of the nozzle is equal to the inner shell one. The nozzle outlet is equipped with the levelling grids and three sensors to measure an air temperature at the bundle inlet. The nozzle is covered with a thermal insulation layer to prevent heat fluxes from electric current supplies to the air flowing through the nozzle. Then the air enters a helical tube bundle, removes heat released in the heated central zone composed of 37 tubes and is exhausted into atmosphere.

Heat is supplied to the tubes from the d.c. ~90 kW generator. The direct current avoids power induction in the metal elements of the design of the experimental section. A maximum current strength during long heat load is 5000 A at a voltage of 18 V. A generator voltage is regulated by varying excitation circuit current. In addition, the energy release power in the heated zone of a tube bundle is regulated, too. A special election device keeps voltage stabilization on the generator terminals. As a result, a constant voltage drop is maintained in a tube bundle under steady conditions. A current strength of 2000 A is measured by a voltage drop on the shunt with a rating accuracy of 0.5. A block of prescribing voltages which can cause sharp time variations of energy release in the heated tubes is mounted to implement unsteady heating of a tube bundle in the generator excitation circuit.

A schematic of the experimental section equipped with a 127-helical tube bundle is shown in Fig. 2.2. A bundle is located in a hexagonal shell. Heated helical tubes were chosen as to have the same wall thickness by measuring their resistance. This provided uniform heat release over the radius of a heated zone of a bundle. Copper tips were welded with silver solder to the bottom ends of steel Kh18N10T tubes. Nickel tips were welded to the top tube ends heated to high temperatures. These tips were connected with electric current busbars. An outlet bundle temperature was measured by 10 chromel-alumel thermocouples embedded at the cell centers with the following values of a relative radius r/r_{sh} = 0.073, 0.128, 0.193, 0.265, 0.334, 0.408, 0.479, 0.624, 0.770, 0.916 and located on the traverse gear.

HEAT CARRIER MIXING **49**

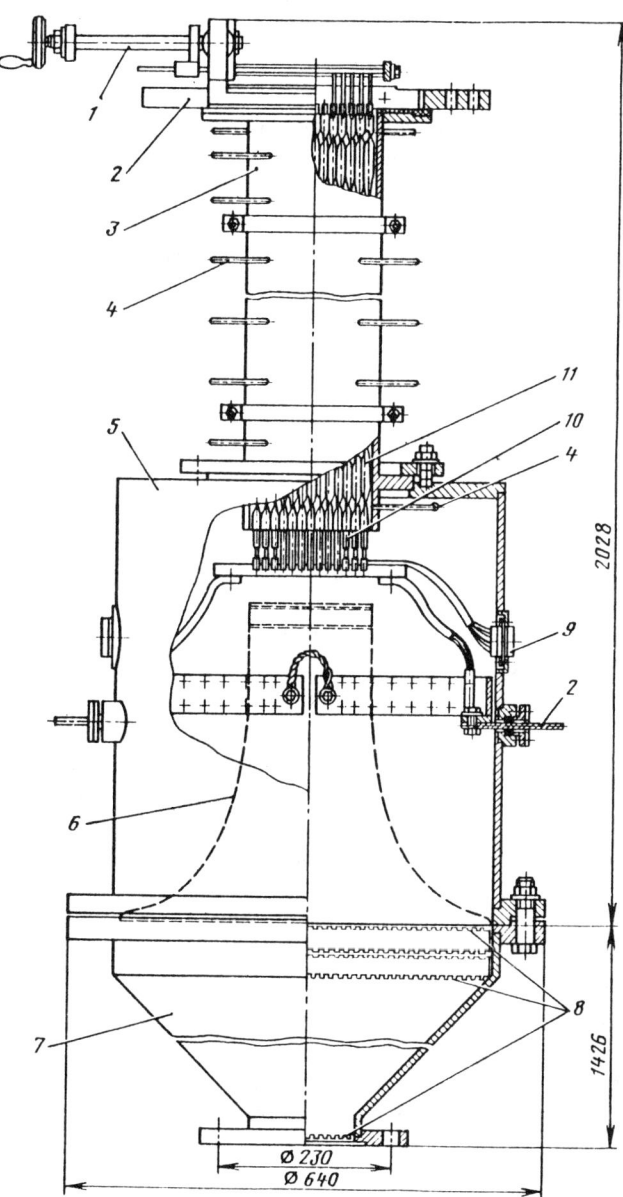

Figure 2.2 Experimental section: *1*) traverse gear; *2*) electric current supplies; *3*) shell; *4*) static pressure pick-up; *5*) inlet chamber; *6*) nozzle; *7*) inlet connection; *8*) grid; *9*) plug connector assembly; *10*) electric current supplying pipe; *11*) helical tube.

Velocity heads were measured by the Pitot tubes 1 mm in dia and 0.1 mm in wall thickness and by the inductive differential pick-ups.

When the heat power is varied in time, a constant heat carrier flow rate via a bundle is kept by a limiting throttling orifice, in which a supercritical pressure drop is set in air flow.

Oval helical tubes were 0.5 m long, a maximum size of an oval was equal to $d = 12.3$ mm and a wall thickness amounted to 0.2 mm. Unsteady heat and mass transfer was studied in tube bundles with relative twisting pitches $S = 12\ d(\mathrm{Fr_M} = S^2/dd_{eq} = 220)$ and $S/d = 6.1$ ($\mathrm{Fr_M} = 57$). In the outlet cross-section of a bundle the major axes of the tube ovals were parallel and the tubes formed the slotted channels for heat carrier flow.

The following procedure was employed to make experiments on unsteady heat and mass transfer at a time variation of heat power. A certain air flow rate was set. Two values of heat power (in terms of relative values of a maximum generator power), at which an unsteady process was implemented, were assigned at the power regulator. During this transient process measurements were made of heat carrier temperature fields at the inlet and outlet of a tube bundle as well as of a bundle voltage drop and current strength in the heated zone of a bundle. Experiment was controlled and parameters were measured automatically by pressing a button "START" of the apparatus described in the next section.

Figure 2.3 shows a 37-tube experimental section. Unsteady temperature fields were studied at the experimental section outlet with a time variation of heat power at heating of all the helical tubes of a bundle. Experiments were made in a bundle 1 m long with $S/d = 12.2$. A thickness of tube walls was 0.5 mm, equivalent bundle diameter $d_{eq} = 7.39$ mm and bundle porosity $m = 0.52$. A corrosion-resistant steel shell had a longitudinal connector assembly, which was sealed by placing a silk thread impregnated with heat-resistant varnish. The inner shell side was coated with aluminum oxide to electrically insulate the tube bundle from the shell. Static pressure holes were made in the shell at distances of 0.35 and 0.75 m from the bundle inlet. A corrugated membrane hindering air leakage into the space between the shell and the casing was welded to the bottom part of the shell to compensate thermal expansion of the latter. The space between the shell and the casing was filled with glass-fibrous heat-insulating material. Helical tubes were fastened to current leads almost in the same manner as in the case of mounting the helical tubes in the experimental section shown in Fig. 2.2. The probes between the current lead and the output connection pipe were placed at the bundle outlet to measure velocities and temperatures. The tubes were oriented in a bundle as in the case of the tube arrangement in the set-up in Fig. 2.2. The thermocouples used for measuring a wall temperature were welded to the inner surface in seven tubes of a bundle at distances of 0.04, 0.072, 0.130, 0.210, 0.350, 0.540, 0.7 and 0.8 m from the bundle

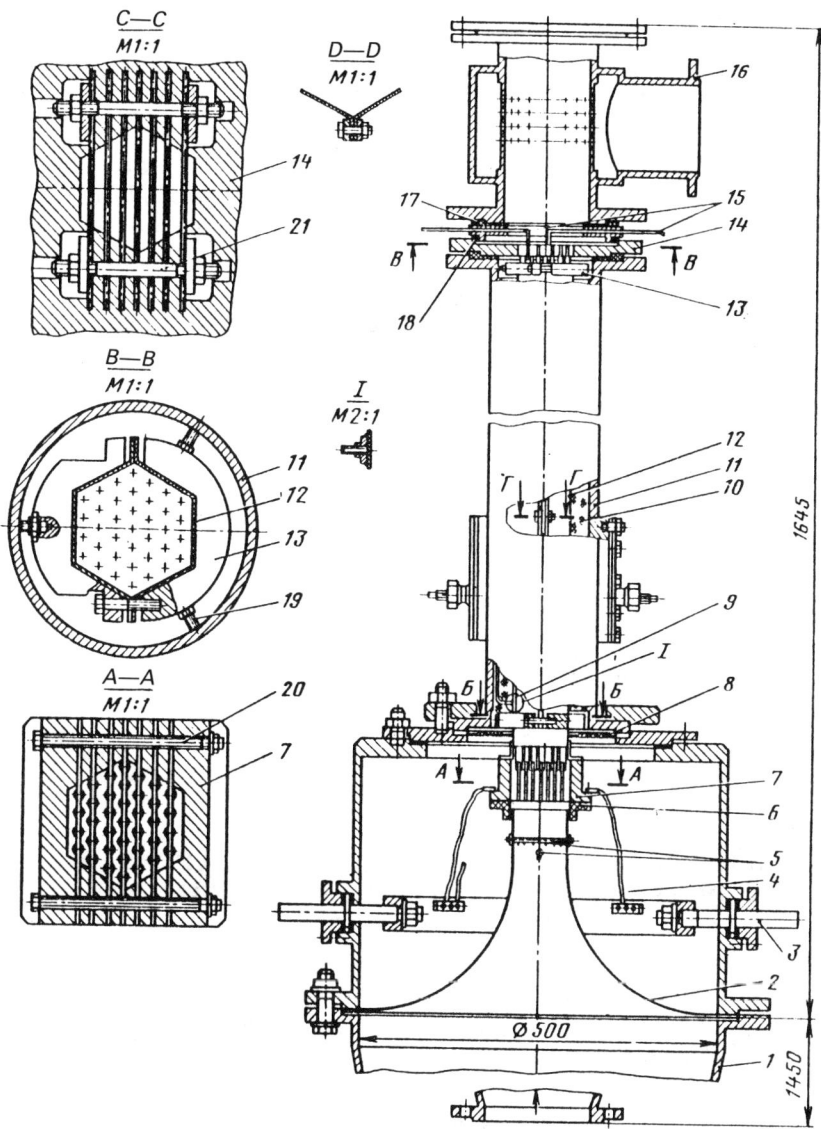

Figure 2.3 Experimental section consisting of 37 tubes: *1*) diffusor; *2*) inlet profile; *3*) electric current supply; *4*) flexible busbars; *5*) thermocouples at the bundle inlet; *6*) intermediate ring; *7*) unit of electric current supply to tubes; *8*) sealing corrugated plate; *9*) static pressure holes; *10*) asbestos-concrete slab (or air sublayer); *11*) shell; *12*) casing; *13*) clamping and centering ring; *14*) top electric current supply and remote-acting grid; *15*) velocity transducer and temperature-sensitive element; *16*) outlet connection; *17, 18*) top sealing rings; *19*) centering fixing rods; *20, 21*) coupling bolts.

inlet. A tube bundle was heated by direct current of the generator. A special electronic device provides an exponential time variation of heat power.

Experimental study of steady heat carrier mixing in a twisted bundle of helical tubes by the method of heat diffusion from a source continuously emitting heated air was made on the experimental set-up shown in Fig. 2.4. Four rows of helical tubes *1* were twisted around a central helical tube in different fashions: γ = const or S = const. A tube bundle was placed into cylindrical shell *4* consisting of four cylindrical sections with flanges. The adjacent sections were separated by the circular organic-glass inserts, in which four holes were made to measure a static pressure. Sections I, II and III were at distances of 375, 750 and 1125 m from the bundle inlet, respectively, downstream. In these sections, a flow temperature was measured by copper-constantan thermocouples which were welded to the inner side of the tubes. About 50 thermocouples were located in one section. 13 thermocouples were mounted on the symmetry line across the inlet cross-section at the bundle inlet. At the bundle outlet a temperature was measured by the thermocouples mounted on traverse gear *2*.

A shell was set up on inlet chamber *5* provided with inlet profile *6* for smooth jet compression and with heated air tube *8*, and thermocouples *7* were led out via the inlet chamber.

A design maintained the heated air tube (inner tube diameter was 34 mm) travel across the section. Experiments were made at three locations of a hot jet. Air was heated approximately up to 400 K, and its flow rate was about 7% of the cold air one. In experiment, the velocities of the hot jet and the main flow were the same.

2.3 AUTOMATIC EXPERIMENT CONTROL APPARATA AND A MEASURING SYSTEM FOR STUDYING UNSTEADY HEAT AND MASS TRANSFER

The instrumentation used to investigate steady mixing of heat carrier at non-uniform heat supply over the bundle radius is unfit for studies of transient heat and mass transfer processes occurring for several seconds. Only a special automatic system may meet the requirements for quick-response and small inertia of control and measuring systems in this case. Therefore, an automatic system (Fig. 2.5) consisting of a measuring-computational complex UVK-2, d.c. generator, pressure transducer, generator power regulator and information converter has been developed to acquisit and process data on unsteady heat transfer and mixing. When a start pulse is supplied from the information converter, the power regulator exponentially changes the output generator power with a prescribed constant of time and voltage polarity within the set limits. Temperatures, pressures and voltages measured in this case and converted into an electric signal are normalized by ther-

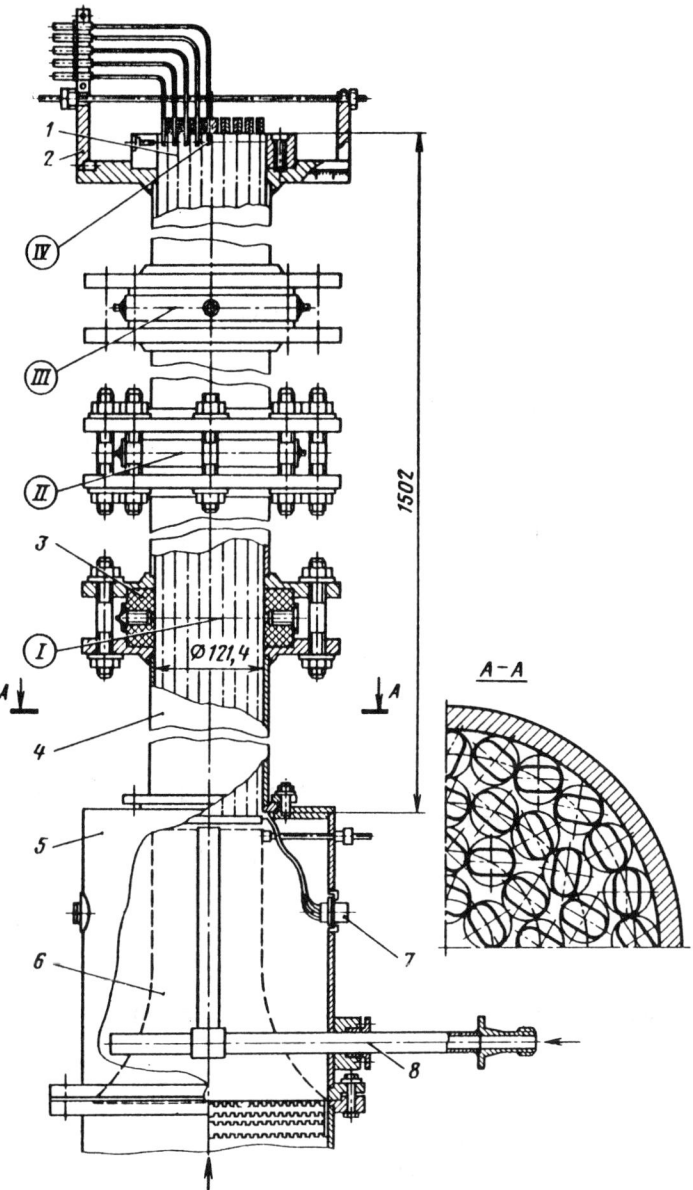

Figure 2.4 Experimental set-up for studying mixing in a twisted bundle of helical tubes: *1*) twisted helical tubes; *2*) traverse gear; *3*) cross-sections for measuring static pressure and temperatures of the flow; *4*) bundle shell; *5*) inlet chamber; *6*) inlet profile; *7*) lead of thermocouple wires; *8*) heated air tube; *I–IV*) sections of location of the thermocouples measuring temperature fields of heat carrier.

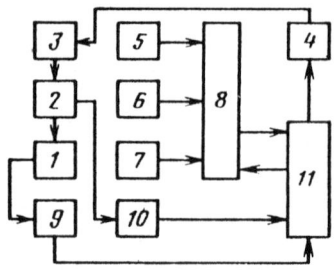

Figure 2.5 Block diagram of the experimental set-up with an automatic system of control and measurement: *1*) experimental section; *2*) shunt; *3*) generator; *4*) power regulator; *5*) thermocouple signal normalizer; *6*) pressure converter; *7*) attenuator; *8*) information converter; *9, 10*) digital voltmeters; *11*) measuring-computational system UVK-2.

mocouple signal amplifiers, a pressure transducer and attenuator, respectively, and the information converter records signal voltages with a polling rate chosen by an operator, converts analog signals into the digital ones and consistently communicates the data into the complex UVK-2 for their further processing and storage. The automatic system enables one to vary the generator power in jumps or exponentially from 0 to 90 kW and, vice versa, as well as according to a special program. All the probes are sampled simultaneously in 40 measuring channels. A time sampling spread is less than 0.7 μs. A sampling rate is from one sampling for 120 s to 25 samplings per second. Digital information is output to a computer with a frequency of 10 kHz. The designed experimental set-up with an automatic system of control, data acquisition and processing enables one to study unsteady mixing in helical tube bundles with sufficient accuracy.

Elements of a system operate as follows. After the necessary limits and time constant of varying the generator power and voltage polarity as well as the start pulse of the information converter are set, the power regulator, whose block diagram is shown in Fig. 2.6, starts varying the generator power. The voltage regulator stabilizes the output generator power with varying the load impedance, programs a time variation of the output power according to the prescribed parameters and checks their programming as well as reverses the generator voltage.

A generator power is specified by a diagram of an analog multiplier based on the time-pulse multiplication principle. Normalized voltages at the generator outlet and on the shunt are supplied to the multiplier. A voltage proportional to the load power is supplied to a comparison circuit drawn up according to an error integrator circuit. A constant voltage determined by the prescribed limits or exponentially varying according to a program circuit is also supplied to the comparison circuit. The programmed power is controlled by a digital voltmeter. The comparison circuit detects a voltage error which gives a gain over the power. A mean excitation winding current is proportional to this error. The reverse of the generator excitation current is initiated manually, if necessary.

Since at zero excitation current the generator has a 0.5 V output voltage, a reverse excitation circuit is stipulated to compensate this voltage. A supply

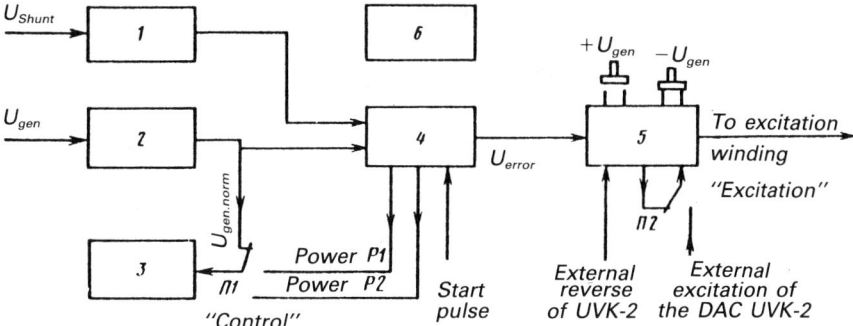

Figure 2.6 Block diagram of the power regulator: *1*) current normalizer; *2*) voltage normalizer (regulated); *3*) control voltmeter; *4*) analog multiplier with a program and comparison circuit; *5*) power amplifier with a reverse circuit; *6*) power source.

source generates a stabilized voltage for energizing an electric circuit. The leading phase feedback provides accomplishing a variation of the power and its stable value at each time moment when a closed regulation circuit is realized. Besides the information converter, the start pulse may be produced by the power regulator itself or by UVK-2.

In studying steady mixing, the voltage regulator is used to stabilize an output generator power.

A thermocouple signal normalizer consists of 20 current amplifiers Φ7025/5 having no galvanic conduction. A thermocouple voltage is supplied to the amplifier input via a screened line. A stable wire-wound resistor serves as amplifier load. The input and the output of the amplifier are equipped with integrating condensers, each having a time constant of 2 ms and the amplifier casings are earthed to escape the influence of electric fields. The current gain is set by a potentiometer. The accuracy class of the amplifier is 0.05. Variable-induction pick-ups with the electronic pressure transducer KWS6A-5 are utilized to measure a pressure. Pressure drops are measured by inductive differential pick-ups PD1 intended for 0.01, 0.1 and 1 atm (0.98, 9.8 and 98 kPa) and a static pressure, by strain gauges P3M intended for the normal range of 10 atm (980 kPa). An error of inductive pick-ups is 10%, of strain gauges is ≤0.25% and of the converter is 0.5%. The inertia of an electric field is ≤0.2 ms. A frequency band of a measured pressure at a level of 3 db is 1.5 kHz.

An attenuator converts amplitudes of measured voltages into the values up to 1 V to change a measuring range before these are input into the information converter. It consists of resistor voltage dividers C5-55 with a division ratio switch: 1, 2, 5, 10 and 20. The accuracy rating of the resistors is 0.1%.

The information converter measures signals in 40 measuring channels, digitizes and transmits them into the information storage or into the computer. It consists of a relay-condenser block of galvanic conduction and signal voltage sampling, noncontacting quick-response commutator of signals, analog-to-digital converter (ADC), output resistor of digital information, output amplifiers, shaper of a microcycle, control block and power supply block (Fig. 2.7). This schematic is chosen from a necessity to have a simple block of galvanic conduction and signal amplitude sampling, the smallest scatter of a measuring time in measuring channels, the largest suppression of the effect of the power line voltage hum and pulse noises of an electric field on a measured signal and on a rapid change of the number of operating measuring channels. High quick-response of a system for data storage was provided by analog keys with field-effect transistors (Fig. 2.8). When the keys operate on overall charge, the total leakage current due to common-mode voltage of the thermocouples attains a value commensurable with a legitimate signal. A schematic of the galvanic conduction of the signal transducers has been adopted to escape the impact of the common-mode voltage on a measuring accuracy of a signal. The galvanic conduction method is based on the capacitors switched by reed relays. Each measuring channel contains its reservoir capacitor and two two-contact relays. In response to a command "START OF MEASUREMENTS" all the capacitors are simultaneously connected with signal transducers and are charged to the voltage level of a signal. All the relay windings are simultaneously de-energized.

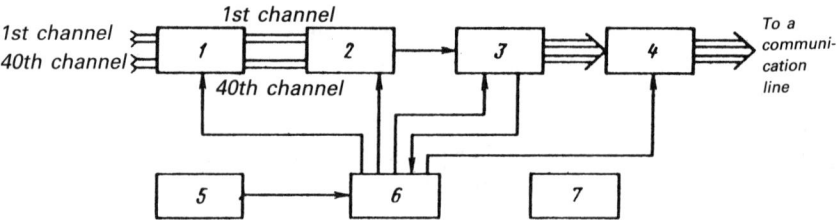

Figure 2.7 Circuit of the information converter: *1*) relay-capacitor galvanic conduction block; *2*) noncontacting commutator; *3*) analog-to-digital converter (ADC); *4*) output register with amplifiers; *5*) shaper of a microcycle; *6*) control block; *7*) power supply block.

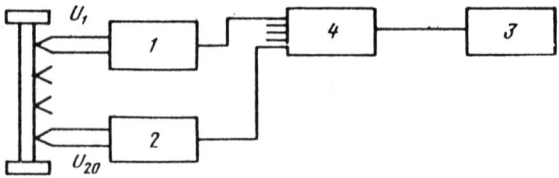

Figure 2.8 Circuit for measuring a thermocouple signal: *1*) normalizing amplifier No. 1; *2*) normalizing amplifier No. 2; *3*) analog-to-digital converter (ADC); *4*) commutator.

The time of the contact fault is ≤0.7 ms. So, the simultaneous information sampling is kept in all 40 channels. The largest suppression of the power supply hum is provided by a scheme referring the charge time and the capacitor switch-off time to the period and phase of the supply line. Use of the capacitors provides common-mode voltage suppression. When a signal transducer goes out of service and several measuring channels are not used, the operation of ADC without load is provided by expelling the unnecessary analog keys of the noncontacting commutator with the aid of the shift register.

All the blocks and circuits of the information converter are controlled by the pulses of the appropriate frequencies and time which are produced by the master generator and the shaper of the frequency pulses of the supply line (Fig. 2.7) in the control block. The master generator produces pulses with a frequency of 200 kHz. These pulses are applied to the shaper of two pulse sequences, i.e. stroke pulse "A" (SP "A") and stroke pulse "B" (SP "B") with frequencies of 100 kHz and shifted half the phase. After a number of divisions are made, the sequence of SP "A"'s is converted into 10 kHz SP "A"'s of the shift register to control the noncontacting commutator and into the pulses to produce a 15 ms microcycle. In the shaper, the position decorder produces 15 ms pulses to form the start of recording (charging of the capacitors)—1st pulse; the end of recording (switching off the capacitors from the measured signal sources)—8th pulse; the start of the counting interval (connecting the capacitors with the noncontacting commutator)—10th pulse; and the start of recording "1" in the shift register (start of the noncontacting commutator for alternate connection with the analog-digital converter)—14th pulse. The end of the counting pulse is formed by a demand pulse of the last capacitor. The stroke pulses having a frequency of 100 kHz and SP "B" sequence (Fig. 2.7) are used to start ADC.

A trigger count-down circuit is intended for producing cycle triggering pulses with a required frequency. Experiments with unsteady mixing of heat carrier were made with frequencies of 50, 20, 10, 5 and 1 Hz.

A cycle counter in the control block (Fig. 2.7) is intended for producing a reset pulse after one or hundred cycles of operation of the information converter. The operation cycle of the information converter starts with pressing a button "START" when one pulse with certain parameters is formed and enters the synchronizer of locking in supply line frequency. The synchronizer initiates a pulse entering the synchronizer of locking in fundamental frequency and appearing simultaneously with the first pulse formed by a pulse shaper of the supply line frequency. The synchronizer of locking in fundamental frequency initiates a pulse, changes over the trigger "OPERATION" to state 1 and sets all the device assemblies into the initial state. In changing over the trigger, a recorder operation clear signal is sent and then a microcycle shaper starts operating. After the above pulse operations are made, the recording 8 ms interval is formed by the trigger "RECORD-

ING" and the counting 3+ (0.1–4) ms interval, by the trigger "COUNTING" and the information interval, by the trigger "INFORMATION". A circuit "I" initiates ADC start pulses under the action both of a high level of the pulse of the trigger "INFORMATION" and of stroke pulses of the SP "B" sequence.

In connecting a measured signal with ADC by means of the noncontacting commutator and in sending an ADC start signal, a measured quantity is digitized. After an ADC signal "MEASUREMENT END" is detected, the digital code is recorded in the output register and then transmitted into a computer via the output amplifiers. After the last measuring channel is sampled and information is transmitted into the computer, the microcycle ceases at the cycle counter position: "1 cycle" or is repeated 100 times at the position: "100 cycles".

A measuring-computational complex (UVK-2) is used for primary data processing and acquisition (Fig. 2.9). The blocks of UVK-2 are connected by a common busbar. The complex UVK-2 is connected via the "CAMAK" device with the external objects. The complete set of the complex contains two CAMAK crates. The block-diagram of the crate is shown in Fig. 2.10. The probe voltages digitized by an information converter (IC) are supplied to the input register. Data are first collected in the on-line storage (CM 3102) to increase a signal detection rate. After a complete cycle (e.g. 40 channels with 100 measurements) is accepted, the collected information is recorded and converted into physical quantities. Depending on the type of a program

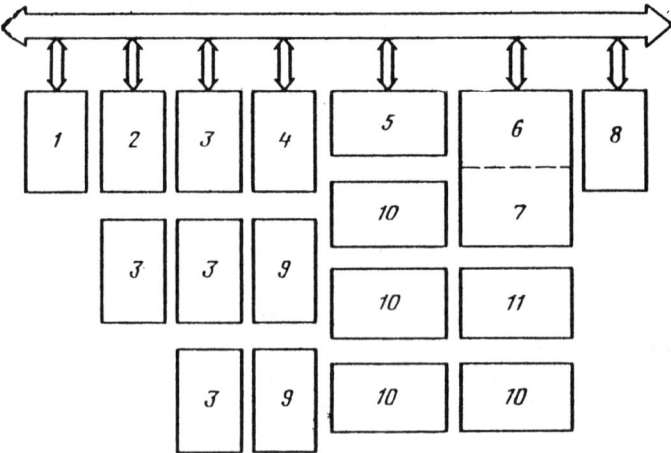

Figure 2.9 Block diagram of the system UVK-2: *1*) processor CM 2104; *2*) device of punched tape input/output, CM 6204; *3*) on-line storage, CM 3102; *4*) controller of magnetic disc storage, CM 5402; *5*) controller of magnetic tape storage, CM 5002; *6*) crate-controller; *7*) CAMAK moduli; *8*) display CM 7204; *9*) magnetic disc IZOT 1370; *10*) magnetic tape storage IZOT 5003; *11*) CAMAK moduli crate.

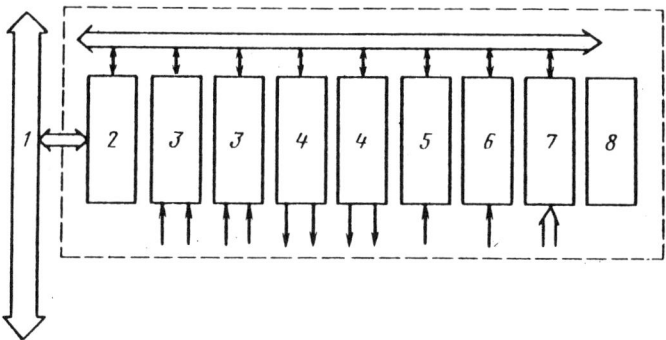

Figure 2.10 Block diagram of the CAMAK crate: *1*) common busbar; *2*) crate controller; *3*) input register, 305; *4*) output register, 350; *5*) ADC 712; *6*) DAC DAC; *7*) relay commutator 750; *8*) supply voltage converter, 0.58.

the data are printed out in the form of tables and charts convenient for further data processing and are stored on the magnetic discs, tapes or punched tapes.

The additional communication of a computer with an object consists of two digital voltmeters (Fig. 2.5) measuring generator voltage and current (shunt voltage). Voltmeter output signals are supplied to the register inputs in a digital form via an interface circuit. A processor calculates a generator power, compares it with the one prescribed according to the program for each time instant and digitizes a voltage error on the digital-analog converter (DAC). A voltage error of the DAC output in an analog form enters the inlet of the external control of the power regulator. The power regulator sets a generator excitation current integrating a voltage error. In varying the heat power, a transient process and information collection may be also controlled by the display via the output CAMAK register.

In studying unsteady mixing of heat carrier in helical tube bundles, the accuracy of the results obtained may be also affected by the inertia of the temperature gauges. Indeed, if in measuring stationary temperature the measuring errors appear because of heat removal from a temperature gauge by conduction due to radiative heat transfer with the surrounding bodies and for some other reasons, then in measuring a time-varying temperature extra errors arise due to the unsteadiness of a process. This is due to the fact that a thermocouple bead does not succeed in taking up an ambient temperature instantaneously, and a signal initiated in a temperature-sensitive element is recorded with a delay because of the thermal inertia of this element. The available methods of calculating the thermocouple inertia enable one to make only approximate estimates of a measuring error of a heat carrier-air temperature. With increasing a heat transfer coefficient as well as with decreasing a thermocouple bead diameter (wire thickness) the inertia decreases. The accuracy of the measuring error may be also affected by a heating rate of

60 UNSTEADY HEAT AND MASS TRANSFER IN HELICAL TUBE BUNDLES

a helical tube bundle or by a time derivative of a heat carrier temperature.

In studying unsteady heat and mass transfer, it is enough to set a time constant of the transient process to estimate a non-stationary error. It will be considered that a temperature difference between a chromel-alumel thermocouple T_{ther} and air flow T_f is equal to $(T_f - T_{ther})$, and an absolute measuring error Δ is the same for steady and unsteady heating processes. Then the time of setting the device indications by means of this converter is

$$\tau_1 = \tau_0 \ln \frac{T_w - T_f}{\Delta} \qquad (2.11)$$

where τ is the time constant. If $\tau_1 = 3\tau_0$, then $\Delta = 0.05 (T_f - T_{ther})$, i.e. the thermocouple bead temperature differs from the ambient one by 5%. At $\tau_1 = 5\tau_0$ $\Delta = 0.007 (T_f - T_{ther})$, i.e. the difference is 0.7%. According to the State Standards for industrial thermocouples an estimate of the thermal inertia is given in time, for which the device indications in measuring an ambient temperature within certain limits do not approach the values corresponding to this temperature by a value which is greater as against a device error. Hence, this condition is satisfied with a thermocouple, whose non-stationary error does not exceed 1% with regard to the thermal inertia. As a result of this estimation of the thermocouple inertia the time constant τ_0 as a function of thermocouple electrode diameter (Fig. 2.11) is obtained. As seen from Fig. 2.11, at a thermocouple electrode diameter $d = 0.2$ mm $\tau_0 = 0.0026$ s. The quantity τ_0 vs the flow velocity ranges $\tau_0 = 0.0026$–0.008 s. At a sharp change of a temperature of air flowing past the thermocouples mounted on the traverse gear (Fig. 2.2) up to 400 K, the measuring error due to the thermocouple inertia will be greater than 5 K in time $\tau_1 = 4.5 \tau_0$ and at a 1000 K temperature jump the measuring error due to the thermocouple inertia will be also greater than 5 K in time $\tau_1 = 5.5 \tau_0$. Then for the thermocouples with $d = 0.2$ mm their inertia will not exceed 0.04–0.2 s, and a non-stationary error will not exceed an absolute measuring error by the device one.

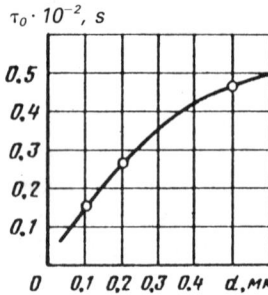

Figure 2.11 Time constant vs. thermocouple electrode diameter.

Thus, the experimental set-ups and the experimental sections provided with an automatic system for data acquisition and processing and for experiment control enable one to study unsteady heat and mass transfer processes with sufficient accuracy.

Specific Features of Unsteady Mixing of Heat Carrier Due to its Time-Varying Flow Rate

A special equipment was developed and the inertia of a measuring system for a heat carrier flow rate was estimated to experimentally investigate unsteady mixing of heat carrier, with its flow rate being varied in time. A time variation of a heat carrier (air) flow rate on the experimental set-up was achieved by changing an area of the flow cross-section of a pipeline. A device intended for changing a cross-sectional flow area of the pipeline was placed in front of a standard nozzle measuring an air flow rate. Such nozzles are usually used to measure a gas flow rate and are mounted in the pipelines no less than 50 mm in dia. In these experiments, air was supplied to a tube bundle via a pipeline 150 mm in dia. The error of the air flow rate with respect to a nozzle pressure drop with regard to the disturbances made by placing this device in front of the nozzle did not exceed 1.5%. A schematic of this device for a sharp change of the air flow is shown in Fig. 2.12, and a block diagram of the set-up equipped with this device is presented in Fig. 2.13.

Operation of the device for changing a cross-sectional flow area of a pipeline (Fig. 2.12) is identical to that of the camera stop and proceeds as follows. An electric signal of the control system (based on the computer CM-4) via an amplifier enters an electromagnetic gate valve of a pneumatic distributor, to which the compressed 0.6 MPa air is supplied. Depending on the investigated unsteady process (decrease or increase in flow rate), the compressed air of the pneumatic distributor enters one of two pneumatic chambers equipped with a spring power accumulator of the type 20-20. In addition, the rod of the chamber travels and applies, via a lever, a force to turn the casing, thus setting in motion the tabs which decrease or increase an area of the flow cross-section of a pipeline.

Transfer of disturbances in the form of ambient pressure or velocity variations from a section to a section of the pipeline, through which air is flowing, occurs with some delay. However, as in the air the disturbance waves are propagating with a sonic velocity, the unsteady gasdynamic processes in a tube damp with time that can be compared with the one of a disturbance wave.

The inertia of a measuring system of flow rate may be estimated, considering that according to the program the experiment is controlled by the computer CM-4, whose signal is supplied for varying flow rate and for recording data. From experiment, the delay time between the signal supply

62 UNSTEADY HEAT AND MASS TRANSFER IN HELICAL TUBE BUNDLES

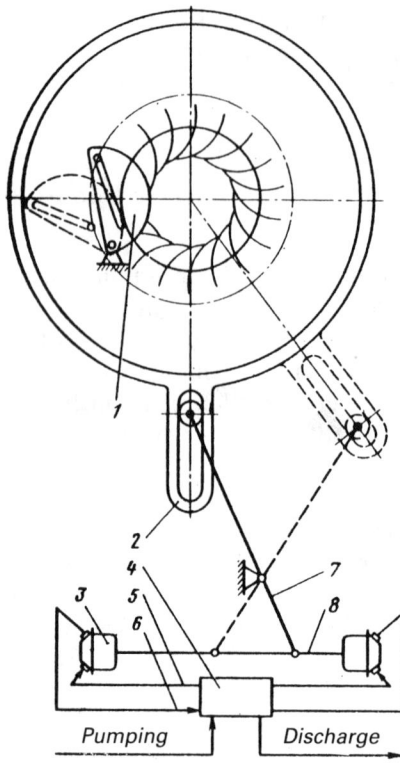

Figure 2.12 Schematic of the device varying a flow rate: *1*) tabs; *2*) shell of the device; *3*) pneumatic chamber; *4*) air distributor; *5*) line of compressed air supply to the pneumatic chamber; *6*) line of compressed air removal; *7*) lever; *8*) rod.

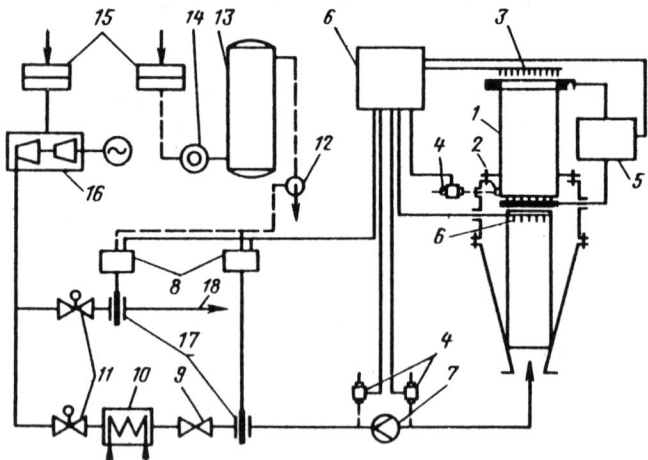

Figure 2.13 Block diagram of the experimental set-up: *1*) experimental section; *2*) pressure holes; *3*) traverse gear with thermocouples; *4*) pressure pick-ups; *5*) d.c. generator; *6*) thermocouples at the bundle inlet; *7*) nozzle for measuring flow rate; *8*) controlling pneumatic activator; *9*) gate valve; *10*) cooler; *11*) automatic-regulated gate valves; *12*) reducer; *13*) receiver; *14*) compressor; *15*) filter; *16*) turbo-compressor; *17*) device for measuring flow rate.

and the device operation was found and did not exceed 0.2 s. Inertia characteristics of the pneumatic pipeline-pressure pickup system were also determined experimentally and compared with the data [45]. It has appeared that the signal delay is no more than 0.04 s. Then with regard to the delay time of 0.03 s in the electronic pressure converter the total delay time of the computer CM-4 signal supplied for the device operation up to the moment of fixing a measured quantity by the device will be no more than 0.1 s. Therefore, the measurements of the parameters for an unsteady process fixed in 0.2–0.4 s at the moment of the command supplied for varying air flow rate are consistent with a gasdynamically steady process in the pipeline at a new value of a flow rate.

CHAPTER
THREE

THE VORTEX FLOW STRUCTURE AND PHYSICAL MECHANISM OF HEAT AND MASS TRANSFER ENHANCEMENT

3.1 THE STEADY-STATE FLOW STRUCTURE IN A HELICAL TUBE BUNDLE

A turbulent flow structure in an oval helical tube bundle is complex in nature. Turbulence in this case is generated both by the nonuniformity of a velocity field near a tube wall and by that of a velocity distribution in the flow core [3]. Therefore, if the turbulence on the axis of a straight round tube is close to homogeneous and isotropic (according to Laufer $\sqrt{\overline{u'^2}}/\sqrt{\overline{v'^2}} = 1.05$), then in a helical tube the anisotropy of properties also manifests itself in the flow core. The nonuniformity of a velocity field in the flow core of a bundle is attributed to the presence of through and spiral channels as well as of the places of contacting the tubes [39]. This may result, under certain flow conditions, in a difference in the vortex structure in these typical flow regions.

Study of the distributions of the relative longitudinal pulsational velocity $\sqrt{\overline{u'^2}}/u_{max}$ in the bundle cross-section at $Re = u d_{eq}/\nu = 6 \cdot 10^3 - 1.1 \cdot 10^5$ and at $Fr_M = 178-1187$ in the typical regions of the flow core (in a through channel and behind a place of contacting the tubes) has shown [12] that maximum values of $\sqrt{\overline{u'^2}}/u_{max}$ are observed behind the points of contacting the adjacent tubes and minimum ones, in through channels. A mean value of the quantity $\sqrt{\overline{u'^2}}/u_{max}$ in the flow core is determined from the expression

$$\left(\frac{\sqrt{\overline{u'^2}}}{u_{max}}\right)_{mean} = 7.2 \text{Re}^{-(0.155 + 40.57 \text{Fr}_M^{-1} + 1700 \text{Fr}_M^{-2})}$$

$$\times \left[1 + \frac{\text{Fr}_M - 178}{7.5(19.5 - 0.135 \text{Fr}_M)}\right] \quad (3.1)$$

Relation (3.1) is also consistent with the experimental turbulence intensity data [39] obtained by the method of heat diffusion from a point source and based on the limiting solution to Taylor's equation at a small diffusion time which is valid both for isotropic and homogeneous turbulence.

Since turbulence anisotropy is usually analyzed depending on vortex sizes, in [12] use was made of the spectral distributions of the longitudinal pulsational velocity component

$$\overline{u'^2} = \int_0^\infty E(f) \, df \quad (3.2)$$

In studies of a contribution of different-scale vortices to energy transfer in the flow, it is found that turbulence in a helical tube bundle contains both large and small energy vortices. As energy dissipation due to viscosity increases with decreasing a vortex size, the energy turbulence spectrum observed in a helical tube bundle moves up in frequency, as compared to the one in a round tube [12], and is responsible for increasing hydraulic resistance as against the one in a round tube. Expressing the quantity $\overline{u'^2}$ in the form of a spectrum through the wave numbers

$$\overline{u'^2} = \int_0^\infty E(k) \, dk \quad (3.3)$$

where

$$k = 2\pi f / u \quad (3.4)$$

and using Taylor's hypothesis on frozen turbulence, estimate was made of a longitudinal integral spatial turbulence scale [12]

$$L = \frac{\pi}{2\overline{u'^2}} \lim_{k \to 0} E(k) \quad (3.5)$$

which characterizes a mean-statistical vortex size. This scale is different for the considered flow regions and does not depend on the Re number.

However, determination of the quantity L by formula (3.5) enables one to estimate its order alone. Therefore, the spatial scale L may be refined by examining the autocorrelation function $\overline{u'(t)u'(t + \tau)}$ and the time correlation coefficient:

$$R_T = \frac{\overline{u'(t)u'(t+\tau)}}{\overline{u'^2}} \tag{3.6}$$

and by using Taylor's hypothesis that this equation

$$\frac{\partial u'}{\partial t} = -(u_{max})_{mean} \frac{\partial u'}{\partial x} \tag{3.7}$$

is also valid at $(u_{max})_{mean} \geq u'$ for the homogeneous flow with a constant velocity. In this case, the autocorrelation coincides with the spatial one for the considered point and for the one displaced at a distance of $(u_{max})_{mean}\tau$ along the x-axis, and its measurement is more reliable since measurement of the spatial correlation is bound up with the distortion of indications of the second hot-wire anemometer measuring u' downstream the hot-wire anemometer at the first point. Use of Taylor's hypothesis on flow with a cross shift and with a large turbulence level, which is typical of helical tube bundles, may be justified by the fact that consideration is made only of the region where the flow core moves with an approximately constant velocity [39]. Then it is possible to determine experimentally the autocorrelation of the quantity u' at the time delay

$$\overline{u'(t)u'(t+\tau)} \qquad \text{76a}$$

the correlation coefficient R_τ, (3.6), the integral time scale

$$\Gamma = \int_0^\infty R_T d\tau \tag{3.8}$$

combining the results on measurements of pulsational components at two points when the quantity $u'(\tau)$ varies under the action of turbulent pulsations as well as the longitudinal turbulence scale

$$L = (u_{max})_{mean}\Gamma \tag{3.9}$$

Experimental study of a longitudinal velocity pulsation intensity, autocorrelation function, spectral density and turbulence macroscales was made in bundles composed of 37 helical tubes with $d = 36$ mm, $S/d = 12.5$ and 25, $Fr_M = 296$ and 1187 [11].

A narrow-band filter, analog hot-wire equipment and digital signal analyzer "Hewlett-Packard" with intermediate magnetic recording of signals are usually used to make a spectral analysis. Let us examine the experimental data obtained by the analog equipment that enables one to make a more rigorous spectral analysis, as compared to the one adopted in [12]. In experiment, turbulent velocity pulsations were recorded in the form of analog shear pulsations. After anomalies and distortions on the digital analyzer are compensated, the analog signals are digitized in discrete implementation and in other operations. By recording is understood the one of hot-wire anemometer indications during a process. In sampling a process, the sampling time step (sampling interval) is chosen from the condition

$$\Delta\tau = 1/(2f_{max}) \qquad (3.10)$$

where f_{max} is the maximum frequency in the spectrum of a given signal. In analyzing velocity pulsations, the fragment of recording was determined through a volume of a storage block N and a number of the chosen recording fragments q ($N = 512$ and $q = 50$).

Normalized autocorrelation coefficients of a process (3.6) were calculated on the analog equipment and on the digital signal analyzer using the inverse Fourier transformation of the estimates of the appropriate spectral densities. A normalized spectral density of a process was estimated by the rapid Fourier transformation in the following way. For the recording fragment the quantity

$$E_k = \frac{2\Delta\tau}{N} |x_k|^2 \qquad (3.11)$$

where x_k is the Fourier series coefficient

$$x_k = \sum_{n=0}^{N-1} x_n \exp(-j\frac{2\pi k n}{N}) \qquad (3.12)$$

where $x_n = x(n\Delta\tau)$ is the recording ordinate, $k = 0, 1, 2, \ldots, N-1$. Ensemble smoothing of spectral density estimates was made by the formula:

$$E_k = \frac{(E_{k1} + E_{k2} + \ldots + E_{kq})/q}{\overline{u'^2}} \qquad (3.13)$$

where $\overline{u'^2}$ is the dispersion of the entire recording.

The results on energy turbulence spectra in helical tube bundles obtained using the above methods are plotted as a function of wave numbers (3.4) in Fig. 3.1. The obtained results (Fig. 3.1) have enabled one to refine the normalized spectral density estimates [12]. In this run of experiments, a spectrum also moves up in wave number, as compared to the one in a round tube. The effect of the Re number on the distribution $E(k)$ practically does not exhibit, however, some trend in elevating energy vortices is observed with increasing the Re number. The same trend is observed in the case of through channels of a bundle, and energy spectra behind the points of contacting the helical tubes move up in wave number. The effect of the Fr_M number on energy turbulence spectra is similar to that of the Re number, i.e. a fraction of energy vortices increases with Fr_M or S/d both for a through channel and behind the points of contacting the tubes (Fig. 3.1). In all probability, when a spectrum moves in the direction of increasing large wave numbers k (large frequencies), it should be expected that the integral lon-

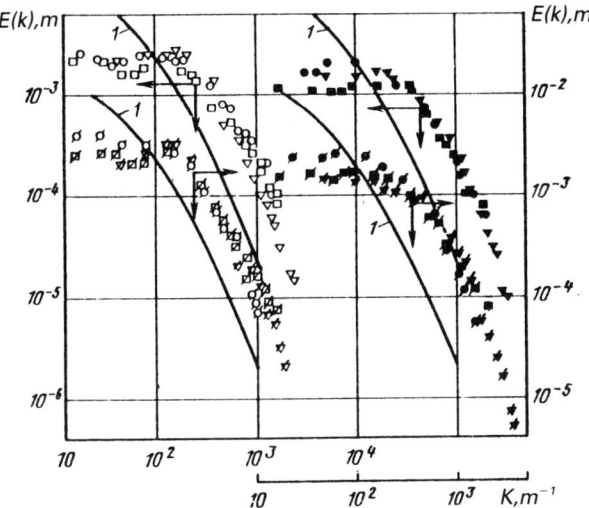

Figure 3.1 Normalized energy turbulance spectra of the quantity u vs wave numbers [11]: ▽, □, ○, ▼, ■, ●, for a bundle with $Fr_M = 296$; ▽̸, ⌀, ⌀, ▼̸, ■̸, ●̸, the same with $Fr_M = 1187$; ▽, □, ○, ▽̸, ⌀, ⌀, for a through channel; ▼, ■, ●, ▼̸, ■̸, ●̸, behind the places of contacting the tubes; ▽, ▼, ▽̸, ▼̸, at $Re = 8300$; □, ■, ⌀, ■̸, at $Re = 2200$; ○, ●, ⌀, ●̸, at $Re = 62000$; 1) data on a round tube.

gitudinal scale of coarse vortices determined by formula (3.9) decreases as well as this scale differs from the one estimated by (3.5) and characterizing a mean-statistical vortex size.

The autocorrelation coefficients determined by the formula

$$R_T = \frac{\overline{u'_n(t) u'_n(t + \tau)}}{\sqrt{u'^2_n(t)} \cdot \sqrt{u'^2_n(t + \tau)}} \quad (3.14)$$

are plotted in Fig. 3.2. It is seen that at large Re numbers $Re = 61400$ R_τ as a function of τ for the considered characteristic points of the flow core may be described by one relation while at $Re = 8300$ a difference between the distributions for a through channel and for a flow region behind the points of contacting the tubes is substantial. Such a pattern is observed for helical tube bundles with a relative twisting pitch $S/d = 12.5$ and 25. This points to the fact that with increasing the Re number the flow core turbulence in a helical tube bundle tends to a more isotropic structure.

At large Re numbers the coefficient R_τ also sharply decreases in the vicinity of the point $\tau = 0$, which is bound up with a wide range of turbulent vortex sizes (Fig. 3.2) in the flow core. For bundles with a tube twisting pitch $S/d = 12.5$ these effects manifest themselves to a greater extent than for those with a large twisting pitch $S/d = 25$.

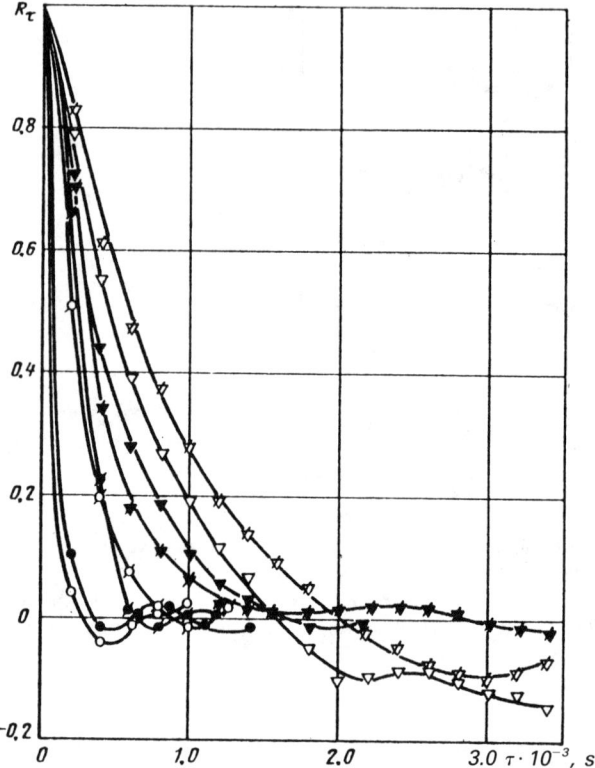

Figure 3.2 Normalized autocorrelation function [11]: ▽, ▼, ○, ●, for a bundle with $Fr_M = 296$; ⌀, ⍒, ⌀, ⍒, for a bundle with $Fr_M = 1187$; ▽, ○, ⌀, ⌀, for a through channel; ▼, ●, ⍒, ⍒, behind the places of contacting the tubes; ▽, ▼, ⌀, ⍒, at $Re = 8300$; ○, ●, ⌀, ⍒, at $Re = 61400$.

The distributions $R_\tau = f(\tau)$ enable one to calculate integral time turbulence macroscales Γ by formula (3.8). Figure 3.3 plots the scale Γ as a function of the Re and Fr_M numbers for the characteristic points of the flow core in the considered bundles. It is seen that for a bundle with $Fr_M = 296$ the time scale in the bundle cross-section in the flow core varies inconsiderably as in the case of a bundle with $Fr_M = 1187$ while, when Re increases from 8300 to 61400, Γ practically decreases by an order of magnitude. Γ as a function of the Re number may be represented by the power functions: for $Fr_M = 296$

$$\Gamma = 4.05/Re^{1.013} \tag{3.15}$$

and for $Fr_M = 1187$

$$\Gamma = 5.26/Re^{1.013} \tag{3.16}$$

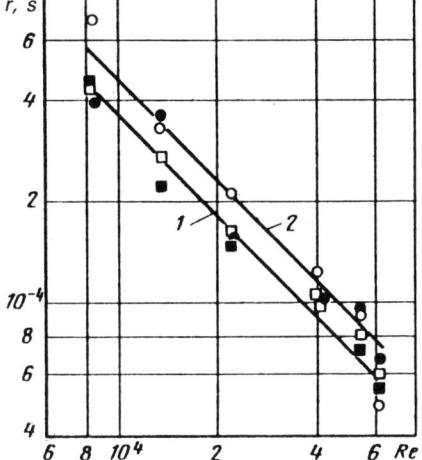

Figure 3.3 Effect of the Re, Fr_M numbers and the hot-wire anemometer location on an integral time turbulence macroscale [11]: □, ■, for a bundle with $Fr_M = 296$; ○, ●, for a bundle with $Fr_M = 1187$; □, ○, for a through channel; ■, ●, behind the places of contacting the tubes; *1*) relation (3.15); *2*) relation (3.16).

Spatial integral turbulence macroscales in a helical tube bundle determined by formula (3.9) may be considered as those of coarse vortices in the mean flow direction. Figure 3.4 plots the experimental data on the scale *L* as a function of the Re and Fr_M numbers for the characteristic points of the flow core. It is seen that the macroscale in the flow core behind the points of contacting the tubes is less than the one in the through channel. Figure 3.4 also plots the mean flow core scale L_{mean} based on the equivalent bundle diameter as a function of the Re and Fr_M numbers. L_{mean}/d_{eq} vs. the Fr_M number is compared, in Fig. 3.5, with the relation from [12]

$$L_{mean}/d_{eq} = 2.41 \cdot Fr_M^{-0.407} \qquad (3.17)$$

It has appeared that for a helical tube bundle with $Fr_M = 296$ the obtained quantity L_{mean}/d_{eq} slightly differs from relation (3.17), and for a bundle with $Fr_M = 1187$ this difference is of the order of 50%. Using the presented data, it may be assumed that the dimensionless parameter L_{mean}/d_{eq} = const (Fr_M) (Fig. 3.5). Then with increasing the Fr_M number a turbulent diffusion contribution to cross mixing of heat carrier in a helical tube bundle will decrease to a lesser degree than according to the estimates [12]. So, for $Fr_M = 296$ this contribution is 10.3–12% instead of 13% [12] and for $Fr_M = 1187$, it is ≈8–9.5% instead of 5.3% [12] under the same assumptions as in [12]. The investigations of the vortex structure of the turbulent flow in helical tube bundles evidence that some anisotropy of properties is observed in the flow core. At large Reynolds numbers the flow structure in this region tends to a more isotropic structure, which is adopted to form the methods of calculating heat and mass transfer in such bundles.

Study of the velocity pulsation intensity, autocorrelation function and spectral density has enabled one to ascertain a physical nature of heat trans-

72 UNSTEADY HEAT AND MASS TRANSFER IN HELICAL TUBE BUNDLES

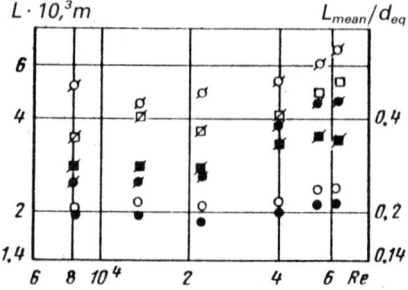

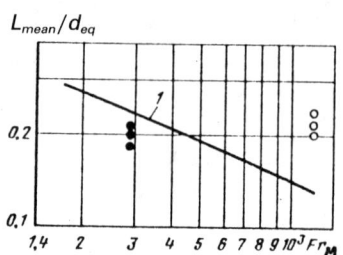

Figure 3.4 Effect of the Re, Fr_M numbers and the hot-wire anemometer location on a longitudinal integral turbulence macroscale [11]: ⌀, ◆, for a bundle with $Fr_M = 1187$; ⌀, ◆, for a bundle with $Fr_M = 1187$; ⌀, ◆, for a bundle with $Fr_M = 296$; ⌀, ⌀, for a through channel; ◆, ◆, behind the places of contacting the tubes; ○, ●, mean values of the dimensionless quantity L_{mean}/d_{eq} at $Fr_M = 1187$ and 296, respectively.

Figure 3.5 Dimensionless scale L_{mean}/d_{eq} vs the Fr_M number: ○, ●, data on $Fr_M = 1187$ and 296, respectively; *1*) relation (3.17).

fer improvement in helical tube bundles. It is proved that additional flow turbulization is bound up with flow swirling and velocity field nonuniformity in the flow core. So, it is observed that an energy turbulence spectrum moves up in frequency (wave number) as against the one in a round tube which is responsible for an energy dissipation increase in the entire flow region and for all the examined Re and Fr_M numbers. In addition, a maximum turbulence intensity is observed in a wake behind the places of contacting the adjacent tubes where an energy spectrum moves, to a greater extent, in the direction of increasing frequencies. An increase in the number of energy vortices with growing the Fr_M number (with growing the relative tube twisting pitch S/d) and a decrease in the turbulence intensity both behind the places of contacting the tubes and in the through channels point to decreasing additional flow turbulization in a helical tube bundle. These laws are also observed in studies of averaged flow characteristics (coefficients of heat transfer and hydraulic resistance) [39].

The fact that the Fr_M and Re numbers do not affect the spatial turbulence scale L_{mean}/d_{eq} (Fig. 3.5) specifying a size of large vortices in the flow direction evidences that in considering heat transfer processes in a bundle d_{eq} may be designated as a scale. A variation of the quantity L in the bundle cross-section (it is less behind the places of contacting the tubes as against the one in the through channel) enables one to support the results on a spectral analysis of the flow structure in a bundle. A turbulent diffusion contribution to interchannel mixing estimated through the measured values of

$\sqrt{\overline{u'^2}}/u_{max}$, L_{mean}/d_{eq} allows one to conclude that ordered convective transfer of heat carrier from one cell of the bundle to another via the spiral tube channels makes a main contribution to this process. The diameter of a helical tube bundle may be, therefore, considered as a scale of the interchannel mixing process. The convective motion in the cross-section of the bundle cells makes a certain contribution to the heat and mass transfer process and promotes intensive transfer of liquid amounts between the wall layer of the tubes and the flow core [3].

3.2 THE UNSTEADY FLOW STRUCTURE AT HEATING AND COOLING

As no data on the unsteady flow structure in helical tube bundles are available, the unsteady turbulent flow structure in a tube is briefly detailed in the present section. The state-of-the art of the problem is reviewed in [26]. Among the recent publications, emphasis should be made on the monographs [5, 33] which are concerned with the velocity profiles and their pulsations at flow acceleration and deceleration in tubes [2, 23, 29, 53, 54, 58], with the temperature fluctuations at varying liquid flow rate [44] and with the unsteady hydraulic resistance coefficients [7, 30].

First, let us consider unsteady isothermal turbulent flow and flow structure in a tube. This problem has been analyzed qualitatively in [26] where for unsteady axisymmetric incompressible liquid tube flow the motion equations, after its terms are comparatively estimated, is written as

$$\rho \frac{\partial u}{\partial \tau} + \rho u \frac{\partial u}{\partial x} + \rho u_r \frac{\partial u}{\partial r} = -\frac{\partial p}{\partial x} + \frac{\tau_x}{r} + \frac{\partial \tau_x}{\partial r} \qquad (3.18)$$

$$\rho \frac{\partial u_r}{\partial \tau} + \rho u_r \frac{\partial u_r}{\partial r} = -\frac{\partial p}{\partial r} + \frac{\rho}{2} \frac{\partial \overline{u_r'^2}}{\partial r} \qquad (3.19)$$

and the continuity equation

$$\frac{\partial u}{\partial x} + \frac{1}{r} \frac{\partial}{\partial r}(r u_r) = 0 \qquad (3.20)$$

where shear stress is

$$\tau_x = (\mu + \mu_T)\frac{\partial u}{\partial r} \qquad (3.21)$$

u_r is the radial velocity component.

In the quasi-stationary case these equations assume the form:

$$\left(\frac{\partial p}{\partial x}\right)_{qs} = \frac{1}{r}\frac{\partial}{\partial r}(r\tau_{x\,qs}); \quad u_r = 0; \quad \frac{\partial u}{\partial x} = 0 \qquad (3.22)$$

In the case of the quasi-stationary motion, a pressure gradient is determined from the one-dimensional motion equation, assuming

$$\left(\frac{\partial p}{\partial x}\right)_{qs} = -\xi \frac{\rho \bar{u}^2}{2d} \tag{3.23}$$

In neglecting the mass forces, we have

$$-\left(\frac{\partial p}{\partial x}\right)_{qs} = -\frac{\partial p}{\partial x} - \rho \frac{\partial u}{\partial \tau} - \rho u \frac{\partial u}{\partial x} \tag{3.24}$$

The pressure gradient is constant over the radius. Hence, from (3.22) we have a linear radius distribution of shear stress

$$\tau_x = \frac{r}{2}\left(\frac{\partial p}{\partial x}\right)_{qs} \tag{3.25}$$

A pressure gradient part spent for shear stress suppression in the unsteady process

$$\left(\frac{\partial p}{\partial x}\right)_\tau = \frac{\tau_x}{r} + \frac{\partial \tau_x}{\partial r} = \frac{1}{r}\frac{\partial}{\partial r}(r\tau_x) \tag{3.26}$$

according to (3.18), is equal to

$$-\left(\frac{\partial p}{\partial x}\right)_\tau = -\frac{\partial p}{\partial x} - \rho \frac{\partial u}{\partial \tau} - \rho u_r \frac{\partial u}{\partial x} - \rho u \frac{\partial u}{\partial r} \tag{3.27}$$

Hence, it is seen that it is variable over the radius.

At flow acceleration ($\partial u/\partial \tau > 0$), the term $-(\partial p/\partial x)_\tau$ monotonically decreases from $-\partial p/\partial x$ on the wall to its minimum $-(\partial p/\partial x)$ on the tube axis. Comparison of (3.24) and (3.27) shows that

$$\left|\left(\frac{\partial p}{\partial x}\right)_{\tau_{min}}\right| < \left|\left(\frac{\partial p}{\partial x}\right)_{qs}\right| < \left|\frac{\partial p}{\partial x}\right| \tag{3.28}$$

A behavior of increasing shear stress modulus with a radius from the axis to the wall

$$|\tau_x(r)| = \left|\frac{1}{r}\int_0^r -r\left(\frac{\partial p}{\partial x}\right)_\tau dr\right| \tag{3.29}$$

is described by the curve sloping up more steeply than $-|\partial p/\partial x|$. At flow acceleration, the shear stress in the flow core is, therefore, less than its quasi-stationary value. This is possible only when the turbulent pulsational intensity $-\overline{\rho u' u_r} = \mu_T \, \partial u/\partial r$ decreases in the flow core. The lag of the turbulent pulsational intensity in the flow core behind their quasi-stationary values may be avoided due to more intensive (as against quasi-stationary) diffusion of kinetic turbulent energy $E_\tau = \overline{\rho u_i u_j}/2$ from the wall region. For this, the generation of E_τ near the wall $-\overline{\rho u'_r u'_r}\, \partial u/\partial r$ should exceed its quasi-stationary value. This is possible if $\sqrt{\overline{u'^2_r}}$ $\overline{\rho u'_r u'_r}$ and $\partial u/\partial r$ simultaneously increase near the wall. But this requires the values of the velocity gradient and shear stress to be higher than their quasi-stationary values near the wall. Figure 1.5 plots qualitative time variations of the shear stress profiles τ_x, turbulent viscosity ϵ_σ and velocity u for accelerated and decelerated flows and their comparison with the quasi-stationary values of $\tau_{x\,qs}$, $\epsilon_{\sigma\,qs}$ and u_{qs}.

The above behavior of velocity, turbulent pulsation and shear stress profiles is consistent with the experimental data of S. B. Markov [23] for a flat channel and of V. I. Bukreev and V. M. Shakhin [2] for a round tube. S. B. Markov showed that in accelerating the flow, a rms value of a longitudinal pulsational velocity component near a wall is higher and in decelerating it, it is less than its stationary distribution, whereas a reverse relationship is valid for a transverse component in the measured region. In the region near the wall turbulent friction at flow acceleration is greater and at flow deceleration it is less than in the case of steady flow.

The unsteadiness very highly affects the hydraulic resistance coefficient in tubes

$$\xi = \frac{\left(|dp/dx| - \rho \dfrac{du_f}{d\tau} 2d\right)}{\rho u_f^2} \qquad (3.30)$$

At flow acceleration, ξ is greater than its quasi-stationary value ξ_{qs} ($\xi/\xi_{qs} > 1$) and at flow deceleration it is less ($\xi/\xi_{qs} < 1$), a difference of ξ/ξ_{qs} from 1 being the greater, the more intensive is flow acceleration or deceleration. However, the quantitative available data considerably disagree. For example, the data of Daily, Hanrey, Olaive and Iordan obtained for water at Re $\leq 5 \cdot 10^5$ and $-0.3 \leq 2a/\xi_{qs} u_f^2\, au_f/d\tau \leq 0.3$ are generalized by the relation

$$K_\xi = \frac{\xi}{\xi_{qs}} = 1 + C \frac{d}{\xi_{qs} u_f^2} \frac{du_f}{d\tau} \qquad (3.31)$$

where $C = 0.01$ at flow acceleration and 0.62 at flow deceleration.

According to the data of A. M. Aitsam, L. L. Paal and W. P. Liiva obtained at the maximum Re number Re $= 16.5 \cdot 10^5$ and $-1.25 \leq d/u_f^2 (du_f/d\tau) \leq 0.8$ the coefficient $C = 1.28$.

A still greater effect of the unsteadiness was revealed in the experiments of I. S. Kochenov and Yu. N. Kuznetsov (down to $\xi < 0$ at flow deceleration) and of S. V. Denisov. In the last work for $K_1 = d/u_f^2\, du_f/d\tau = 0.04$–$0.43$ and $K_2 = 1/\bar{u}\, \sqrt[3]{d^2\, (d^2\bar{u}/d\tau^2)} = 0.12$–$0.86$ at flow acceleration.

$$K_\xi = \exp[-20K_1] + 20K_1 \frac{1+K_2}{1+10K_1} \exp(1+K_1) \qquad (3.32)$$

A considerably less effect of the unsteadiness is revealed according to the data of S. B. Markov [23] and D. N. Popov [33]. S. B. Markov derived the following relations:

$$K_\xi = -0.43 \exp\left(-23 \frac{K_G}{\text{Re}}\right) + 1.43 \qquad (3.33)$$

at $0 < K_G/\text{Re} < 0.214$;

$$K_\xi = \left[12.76 \left(\frac{K_G}{\text{Re}}\right)^2 - 1.38 \frac{K_G}{\text{Re}} + 1\right] \exp\left(3.67 \frac{K_G}{\text{Re}}\right) \qquad (3.34)$$

at $-0.301 \leq K_G/\text{Re} < 0$. In this case, K_G and Re are determined over the flat channel height H. For accelerated flows in a tube $d = 27$ mm in dia and $l = 600$ mm long at $K'_{\text{uns}} = (d/(\xi_{qs} u_f^2))(du_f/d\tau) = 0-80$ and $K_D = (d^2/32\nu)\sqrt{1/u_f |(d^2 u_f)/d\tau^2|} = 0-2000$. D. N. Popov obtained the relation

$$K_\xi = 1 + a_0 K'_{\text{uns}} + a_1[\exp(a_2 + a_3 K'_{\text{uns}}) K_G - 1] \qquad (3.35)$$

where $a_0 = 0.07$, $a_1 = 32$, $a_2 = 0.1 \cdot 10^{-4}$, $a_3 = 0.8 \cdot 10^{-5}$.

For accelerated tube flows N. M. Gulin, G. V. Rineiskaya and L. T. Inozemtseva [7] propose the following relations:

$$K_\xi = 1 + \frac{3.64 d\, du_f}{\xi_{qs} u_f^2 d\tau} \qquad (3.36)$$

for

$$0 \leq \frac{1}{\xi_{qs}} \frac{d}{u_f^2} \frac{du_f}{d\tau} \leq 0.055$$

and

$$K_\xi = 1.2 + \frac{1.4\, d\, du_f}{\xi_{qs} u_f^2\, d\tau} \qquad (3.37)$$

for

$$0.055 \leq \frac{1}{\xi_{qs}} \frac{d}{u_f^2} \frac{du_f}{d\tau} \leq 0.12$$

The investigations made convince that the quasi-stationary method of calculating hydraulic losses is not valid in a general case and does not enable one to reliably establish the limits of the applicability of the quasi-stationary methods because of great quantitative disagreements.

Data on the turbulent structure of unsteady nonisothermal channel flows

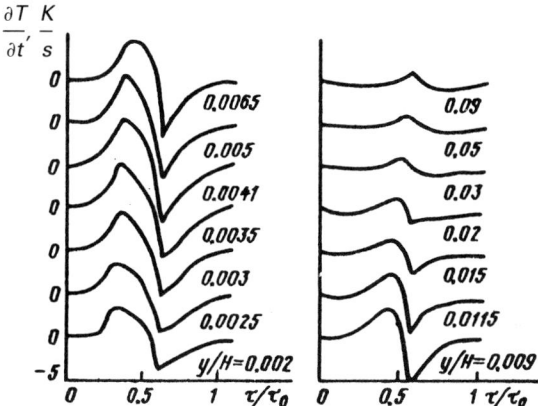

Figure 3.6 Rate of varying a flow temperature at different distances from the wall vs time at Re = 48500 and at a frequency of 0.5 Hz [44]: τ_0, pulsational flow rate period; H, half channel width; y, distance from a wall.

are very scanty. The work of B. V. Perepelitsa, Yu. I. Pshenichnikov, E. M. Khabakhpasheva [44] deals with measurements of statistical characteristics of temperature fluctuations in the unsteady turbulent water flow over the range of Re = $(1.36\text{--}6.1) \cdot 10^4$ and at flow rate pulsations from 0.4 to 4 Hz. Experiments were made in a rectangular channel with its one wall being heated and with a preliminary hydrodynamic stabilization length. The inlet of the experimental section was provided with a pulsar to promote liquid flow rate pulsations. Instantaneous values of the liquid flow rate varied up to 5 times. As heat release in the heated wall in this case did not change, in increasing the liquid flow rate a wall temperature must decrease and in decreasing the liquid flow rate, it must increase. Accordingly, the flow temperature near a wall varies in time. A time variation of the averaged temperature profile is seen from the distribution of a rate of varying the temperature $\partial T/\partial \tau$ during one period. Figure 3.6 plots the quantity $\partial T/\partial \tau$ vs a phase of liquid flow rate pulsations at different distances from a wall. The liquid flow rate via the channel occurs at $\partial T/\partial \tau$ between 0.3 and 0.5–0.6 and increases 0.5 to 0.6 and 1. As seen from this figure, the strongest increase of a temperature is observed in a viscous sublayer region at the start of decreasing the liquid flow rate, and the temperature profile most greatly redevelops in the intermediate zone.

Redeveloping a temperature profile results in a substantial redistribution of the temperature fluctuation level over the channel cross-section. As seen from Fig. 3.7, the flow temperature variation is definitively bound up with that of the liquid flow rate, and a minimum liquid flow rate practically corresponds to a minimum flow temperature at different distances from a wall. In the wall region (determined in terms of a viscous sublayer thickness,

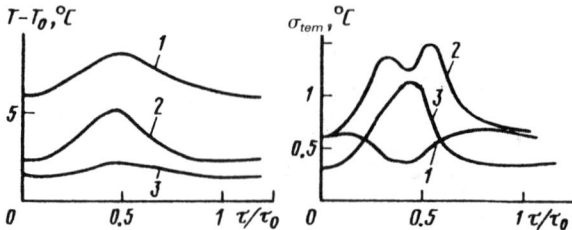

Figure 3.7 Time variation of a local flow temperature and temperature fluctuation rate σ_{tem} at different points of the channel cross-section for Re = 48500 and frequency of 0.5 Hz [44]: 1–3) $y/H = 0.002, 0.009, 0.03$; T_0, flow temperature far from a wall.

using the quasi-stationary relations) a minimum liquid flow rate is consistent with a minimum rate of temperature fluctuations σ_{tem}. In removing far from a wall, σ_{tem} versus time varies, and more intensive temperature fluctuations may be in accord with a minimum flow rate phase. Data on averaged and pulsational velocity component fields as well as on wall shear stress in unsteady nonisothermal flows are needed to generalize experimental results and to work out theoretical models for a phenomenon.

3.3 SPECIFIC FEATURES OF DEVELOPING UNSTEADY TEMPERATURE FIELDS UNDER UNIFORM HEATING OF HELICAL TUBES

Developing unsteady temperature fields at uniform heating of helical tubes was studied in a bundle consisting of 37 helical tubes placed into a hexagonal casing provided with a longitudinal connector assembly (Fig. 2.3). Oval helical tubes were made of corrosion-resistant steel Kh18N10T pipes 10 × 0.5 mm in size. A bundle length was 1000 mm and a relative tube twisting pitch was $S/d = 14.2$.

The casing was made of sheet corrosion-resistant 2 mm thick steel, whose inner side was coated with an aluminum oxide layer and then with a fireproof varnish layer to provide reliable electric insulation of bundle tubes from the casing. The casing was placed into a cylindrical shell and mounted to it by means of a membrane, thus forming a stagnant air layer between them.

In experiment, all the tubes in a bundle were heated. A generator power might be varied according to the prescribed law, and a flow rate of heat carrier (air) remained constant. In experiment, a gas heat carrier temperature was measured by the thermocouples at the centres of the intertube cells. Relative radii of embedding the thermocouples are: $r/r_{sh} = 0.417, 0.567, 0.688, 0.801, 0.847$ where r_{sh} is the radius of a circle whose area is equal to the cross-sectional area of the hexagonal casing.

In changing from one power to another, a new thermal regime sets in not at once but with some delay which is due to different reasons. Let us consider the factors affecting this process. Since heat energy is transferred in a solid (helical tube bundle)-heat carrier system, one has to take into account a heat propagation velocity in helical tubes under unsteady conditions. In this case, consideration is made of homogeneous helical tubes with uniform heat release over the cross-section, whose thermal inertia properties are specified by thermal diffusivity. In considering heat propagation over the helical tube wall, it is possible to adopt the concept of an internal thermal delay time τ_i. This delay time τ_i depends not only on thermal diffusivity determined in terms of the values of thermal conductivity, density and heat capacity but also on geometrical sizes and shape of a body. Thus, a heat energy transfer velocity is defined by the properties of a system consisting of a helical tube bundle (solid).

Usually cooling or heating of a body is intricate. Especially, this is concerned with the initial stage of a process, before a regular regime is set in, in which a body temperature varies exponentially.

The start of a regular heat transfer regime in systems with energy sources occurs considerably faster than in the case of usual heating or cooling of a system in a constant temperature medium. The internal relaxation time τ_{pi} (when a temperature in a regular regime varies e times) may be used to estimate the internal thermal delay time since a difference between these two time characteristics is relatively small and depends on a body geometry, e.g. for a plate the ratio τ_{pi}/τ_i is close to 6/5 and for a cylinder, 4/3 [21].

In a general case, the internal thermal delay time may be estimated by the formula [21]:

$$\tau_i = \bar{\varphi} \, d_g^2 \, \frac{\gamma_{body} C_{tot}}{\lambda_T} \tag{3.38}$$

where d_g is the generalized geometrical characteristic, $\bar{\varphi}$ is the shape factor and γ_{body} is the density.

The shape factor depends neither on properties nor on sizes of a body but it proves to be different for a rod, plate and pipe, $\bar{\varphi} = \varphi \, (R_q^2/R^2)$ where R_q is the radius of the surface, along which cooling takes place, R is the non-cooled surface radius. For a rod we have $R_q^2/R^2 = \infty$, $\bar{\varphi} = 0.0313$ and for a plate with one-sided cooling $R_q^2/R^2 = 1$ and $\bar{\varphi} = 0.0203$. For the ratio $R_q^2/R^2 = 8$ the quantity $\bar{\varphi} = 0.03$ [21].

Estimation of the internal thermal delay time by (3.38) yields $\tau_i = 0.02$ s.

Heat transfer from a helical tube to heat carrier is characterized by the heat transfer coefficient. For one and the same system (of helical tubes) heat energy transfer delay varies over wide ranges depending on a value of the heat transfer coefficient and becomes minimum at $\alpha = \infty$. The time of thermal equilibrium between a helical tube (assuming, infinite heat conduction) and heat carrier is specified by the external delay time τ_e which is equal to [21]

$$\tau_e = \frac{c_{tot}}{\alpha F_q} \tag{3.39}$$

where c_{tot} is the total heat capacity of helical tubes in a bundle and F_q is the heat-releasing surface.

Since from the very beginning this process practically follows the exponential law, the mean cooling time coincides with the relaxation one, i.e. $\tau_{pe} = \tau_e$.

Estimation of the external delay time at Re = $5.5 \cdot 10^4$ and heat carrier flow rate G = 0.441 kg/s yields $\tau_e \approx 5$ s.

The next reason for thermal delay is bound up with limited mass heat capacity of the flow. Indeed, if the first two reasons are excluded, i.e., it is assumed that thermal conductivity of a helical tube and heat transfer coefficient are infinitely large, then the flow cannot instantaneously absorb all heat energy from the channel wall or take it away to the wall. For the time being in contact with the wall the flow may absorb or transfer only a certain amount of heat. The time, for which heat is removed by the heat carrier flow, is referred to as the transport time, τ_{tg}.

A heat balance gives the exponential cooling law, and the mean time of heat removal – transport time of thermal delay is determined by the formula

$$\tau_{tg} = \frac{c + c_{tot}}{Gc} = \frac{l}{w} \frac{c_p}{c}$$

where $c_p = c + c_{tot}$ is the heat capacity of helical tubes and heat carrier. The estimation results in $\tau_{tg} \approx 5$ s.

Figure 3.8a,b plots the experimental results on temperature field formation at uniform heating of helical tubes of a bundle and their comparison with the predicted ones [32]. Experimental and predicted data are compared for two versions in terms of heat power at a constant heat carrier flow rate when heat power increases (Fig. 3.8a) and decreases (Fig. 3.8b).

From Fig. 3.8a,b it is seen that at heating and cooling the experimental and predicted data are in satisfactory agreement. In experiments with bundle heating, a good agreement is observed not only for the experimental and predicted gas temperature fields but also for the time heat carrier temperature derivative $\partial T/\partial \tau$ during a heating process. As for experiments with decreasing heat release, it is seen (Fig. 3.8b) that the cooling time of a helical tube bundle at a constant heat carrier flow rate markedly exceeds the previously estimated transport time of thermal delay. Two reasons are the cause of this phenomenon. The first reason is that the heat flux is thrown off not instantaneously but exponentially and decreases from its nominal one to the quan-

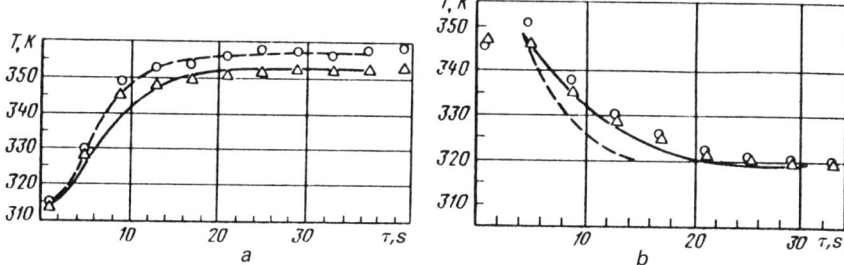

Figure 3.8 Heat carrier temperature in the outlet cross-section of a helical tube bundle with increasing (*a*) and decreasing (*b*) heat power and Re = $5.5 \cdot 10^4$ at different time instants: ○, △, experimental data for relative radii r/r_{circle} = 0.417 and 0.688, respectively; —, with no regard for the mass of current leads; ---, with regard for the mass of current leads.

tity equal to 0.005 for 5 s. The second reason is that the associated masses (current leads to helical tubes, busbars, etc.) are present and may increase a transport time of thermal delay. Calculation results on temperature fields of heat carrier with regard to "associated masses" obtained according to the program [32] satisfactorily agree with the experimental data on decreasing heat power.

This study of unsteady temperature fields at uniform heating of helical tubes of a bundle has enabled one to validate the calculation methods, considering structural specific features of experimental set-ups. The approach elaborated on the grounds of this study to calculate unsteady temperature fields was also employed to generalize the experimental data on nonuniform heat release over a radius of a helical tube bundle, and the obtained results are detailed in Chapter 5.

CHAPTER
FOUR
TRANSPORT PROPERTIES OF FLOW UNDER STEADY CONDITIONS

4.1 HEAT AND MASS TRANSFER IN STRAIGHT HELICAL TUBE BUNDLES

In the case of longitudinal flow past bundles of finned rods and oval helical tubes, the revealed enhancement of interchannel mixing of heat carrier is more considerable as against the case of round tube flow [9, 39, 48]. This is very important for heat exchangers with noticeable nonuniformity of an energy release (heat supply) field in the cross-section of a bundle. Usually temperature distributions in bundles of finned rods are determined by the method of calculating elementary cells with regard to the effects of mass, momentum and energy transfer, and the experimental mixing coefficient $\bar{\mu} = G_{ij}/C_i$ [48] is adopted to close a system of equations. However, in this case, when the number of rods (tubes) in a bundle is great, much computer time is needed to implement a computational program. Therefore, the method of homogenization of a real bundle [39, 9] is used to determine temperature fields of heat carrier in helical tube bundles and is also recommended to calculate temperature fields in the bundles of finned rods.

For the axisymmetric problem, the flow of a homogenized medium with nonuniform energy release and with a temperature- and pressure-dependent density of heat carrier is described by the system of equations (1.8)–(1.11) with boundary conditions (1.12)–(1.14), in which the effective turbulent viscosity coefficient ν_{eff} and thermal conductivity λ_{eff} in (1.8) and

(1.10) may be represented in terms of the effective turbulent diffusion coefficient D_t, assuming that the Lewis (Le) and Prandtl (Pr) numbers are equal to unity:

$$\lambda_{\text{eff}} = D_t \rho c_p \tag{4.1}$$

$$\nu_{\text{eff}} = D_t \tag{4.2}$$

The coefficient D_t for helical tube bundles was determined experimentally in [9]. This coefficient

$$K = D_t/u d_{\text{eq}} \tag{4.3}$$

may be related to the coefficient

$$K = \bar{\mu} p^2 / 4 d_{\text{eq}} \tag{4.4}$$

for the bundles of the finned rods. Here p is the rod grid pitch.

Expression (4.4) is obtained as follows. An amount of heat Q_{ij} transferred from the cell i to the cell j per unit length is equal to:

$$Q_{ij} = G_{ij} c_p (T_i - T_j) \tag{4.5}$$

$$Q_{ij} = \rho h D_t c_p \left(\frac{\partial T}{\partial y}\right)_{ij} \tag{4.6}$$

where $h = p - d_{\text{grid}}$ is the rod grid pitch.

Equating RHSs of expressions (4.5) and (4.6) arrives at $(\partial T/\partial y)_{ij} = (T_i - T_j)/y_{ij}$, $y_{ij} = h$

$$G_{ij} = \rho D_i \tag{4.7}$$

Dividing LHS and RHS of equality (4.7) by

$$G_i = \rho u F_{\text{cell}} \tag{4.8}$$

arrives at

$$\bar{\mu} = G_{ij}/G_i = D_t/u F_{\text{cell}} = K d_{\text{eq}}/F_{\text{cell}} \tag{4.9}$$

Then from (4.9), assuming $F_{\text{cell}} \approx mp^2/2 \approx 0.5\, p^2/2$ yields (4.4).

A criterial relationship between the coefficient K and the main similarity numbers must be established to utilize the homogenization method for calculation of temperature fields in finned rod bundles $m \neq 0.5$.

In the reported works the porosity of finned rod bundles with respect to heat carrier varies over the range of $m = 0.27-0.5$. Therefore, in determining the dimensionless coefficient K through the experimental coefficient $\bar{\mu}$, instead of relation (4.4), the formula

$$K = \bar{\mu} \rho^2 m / 2 d_{\text{eq}} \tag{4.10}$$

should be used, which is obtained from the same considerations as relation (4.4).

Assuming that the flow in finned rod bundles is similar, in its nature, to the one in oval helical tube bundles, it may be said that transport flow properties in them will be almost the same. Then according to [13] a criterial relation for the coefficient K may be sought in the form:

$$K = K(Fr_M, Re, m, x/d_{rod}) \tag{4.11}$$

where $Fr_M = S^2/(dd_{eq})$, $Re = \rho u d_{eq}/\mu_f$.

Let us analyze the publications dealing with interchannel mixing of heat carrier in bundles of finned rods (or rods with spiral arrangement) [17, 18, 42, 48–50, 55, 56], in which use was made both of different methods to experimentally determine the coefficient $\bar{\mu}$ and of various heat carriers. So, works [48, 49, 42] were concerned with use of the method of heating of a central rod ("thermal wake"), [17, 18, 42] with the electromagnetic method (for liquid-metal heat carriers), [50, 55, 56] with the diffusion method based on injection of a more heated heat carrier into one of the cells accompanied by subsequent measurements of flow temperature distributions. In this case, various heat carriers were used: air, water, liquid metals and their compositions (Table 4.1). In generalizing experimental data on mixing, it was assumed that the adopted investigation method and kind of heat carrier do not affect numerical values of the coefficient $\bar{\mu}$ and, hence, the coefficient K which was calculated by (4.10). Geometrical parameters of the examined finned rod bundles, experimental values of the coefficient $\bar{\mu}$ as well as calculation results on the main numbers entering into (4.11) and on the coefficient K are cited in Table 4.1. Mean values of the coefficient $\bar{\mu}$ over the Re number range [17, 18, 42, 48–50, 55, 56] were utilized to generalize mixing data. This is explained by the fact that in the majority of cases, experiments were made either at $Re > 10^4$ or the mean values of Re, at which $\bar{\mu}$ was taken, exceeded $Re = 10^4$. This emphasizes, too that Re does not affect $\bar{\mu}$ in numerous publications. Experiments with helical tube bundles [39] also evidenced that for $Re \geq 10^4$ the coefficient K almost does not contribute to Re, and the impact of Re on K at $Re < 10^4$ is small. For $Re < 10^4$ the coefficient K in helical tube bundles increases with decreasing Re according to the relation [13]

$$K = 3.1623[0.136 Fr_M^{-0.256} + 10 Fr_M^{-0.66} (m - 0.46)] Re^{-0.125} \tag{4.12}$$

In studying a mixing process, the formula [13]:

$$x_{mix}/d_{rod} = 12.2 a^{-1} Fr_M^{-0.275} \tag{4.13}$$

where

$$a = 0.0745 + 11.37 Fr_M^{-1} + 246 Fr_M^{-2} \tag{4.14}$$

was used to estimate the starting length in bundles of finned rods.

The results calculated by formula (4.13) are presented in Table 4.1 which

86 UNSTEADY HEAT AND MASS TRANSFER IN HELICAL TUBE BUNDLES

Table 4.1 Initial geometrical sizes of finned rod bundles. Experimental data on mixing coefficient (bundle nos. 1–3 [48], no. 4 [49], nos. 5–6 [17], no. 7 [18], nos. 8–13 [42], no. 14 [50], no. 15 [55], nos. 16–19 [59]

Parameter	Bundle number					
	1	2	3	4	5	6
Heat carrier	Na	Na	Na	Air	Na—K	Na—K
Twisting pitch, S, mm	100	200	300	100	96	144
d_{cyl}, mm	6	6	6	6	19	19
d_{rod}, mm	9.84	9.84	9.84	8	21	21
S/d_{rod}	10.2	20.3	30.5	12.5	4.57	6.86
d_{eq}, mm	3.96	3.96	3.96	2.58	3.63	3.63
S/d_{eq}	25.3	50.5	75.8	38.8	26.7	40
$Fr_M = S^2/(d_{rod}d_{eq})$	258	1025	2310	485	122	274
$m = F_f/F_\Sigma$	0.476	0.476	0.476	0.36	0.28	0.28
Bundle length, l, mm	1000	1000	1000	660	745	745
p, mm	7.92	7.92	7.92	7.02	20	20
μ, 1/cm	0.108	0.053	0.032	0.067	0.025	0.01
K	0.041	0.020	0.012	0.023	0.036	0.02
Re	$8 \cdot 10^3 \ldots 7 \cdot 10^4$	$8 \cdot 10^3 \ldots 7 \cdot 10^4$	$8 \cdot 10^3 \ldots 7 \cdot 10^4$	$2 \cdot 10^4 \ldots 1.5 \cdot 10^5$	$5 \cdot 10^3 \ldots 2.3 \cdot 10^4$	$2.5 \cdot 10^3 \ldots 2.1 \cdot 10^4$
Number of rods	61	61	61	61	19	19
$x_{St.mix}/d_{rod}$	21.6	21.08	18.1	22.4	17.6	21.8
l/d_{rod}	102	102	102	82.5	35.4	35.4

Table 4.1 continued

Parameter	Bundle number					
	7	8	9	10	11	12
Heat carrier	Na—K	Na—K	Na—K	Na—K	Na—K	Na—K
Twisting pitch S, mm	144	96	144	192	192	375
d_{cyl}, mm	19	19	19	19	16	16.5
d_{rod}, mm	33.8	25.5	25.5	25.5	21.9	21.9
S/d_{rod}	4.26	3.76	5.65	7.53	8.77	17.12

Table 4.1 *continued*

Parameter	Bundle number					
	7	8	9	10	11	12
d_{eq}, mm	12.2	8.47	8.47	8.47	7.45	7.68
S/d_{eq}	11.8	11.3	17	22.7	25.8	48.8
$Fr_M = S^2/(d_{rod}d_{eq})$	50	42.6	96	171	226	836
$m = F_f/F_\Sigma$	0.494	0.39	0.39	0.39	0.39	0.39
Bundle length, l, mm	745	1000	1000	1000	1824	701
p, mm	26.4	22.23	22.23	22.23	18.96	19.55
$\bar{\mu}$, 1/cm	0.073	0.092	0.055	0.048	0.047	0.024
K	0.103	0.105	0.063	0.055	0.044	0.023
Re	$4 \cdot 10^3 \ldots 3.4 \cdot 10^4$	$2.8 \cdot 10^4 \ldots 4 \cdot 10^4$	$4 \cdot 10^3 \ldots 3.5 \cdot 10^4$	$2.8 \cdot 10^4 \ldots 4 \cdot 10^4$	$6 \cdot 10^3 \ldots 5 \cdot 10^4$	$2 \cdot 10^3 \ldots 5 \cdot 10^4$
Number of rods	19	19	19	19	37	37
$x_{St.mix}/d_{rod}$	10.4	9.1	15.8	20	21.1	21.5
l/d_{rod}	22	39	39	39	83	32

Table 4.1 *continued*

Parameter	Bundle number						
	13	14	15	16	17	18	19
Heat carrier	Na—K	H_2O	H_2O	H_2O	H_2O	H_2O	H_2O
Twisting pitch, S, mm	100	300	255	300	450	575	450
d_{cyl}, mm	6.1	5.84	6.3	21	21	21	23
d_{rod}, mm	7.93	8.76	9.07	26.9	26.9	26.9	25.8
S/d_{rod}	12.6	34.3	28.1	11.2	16.74	21.4	17.5
d_{eq}, mm	2.41	3.2	3.35	8.3	8.3	8.3	5.64
S/d_{eq}	41.5	93.8	76.1	36.3	54.4	69.5	79.8
$Fr_M = S^2/(d_{rod}d_{eq})$	523	3210	2140	405	911	1487	1394
$m = F_f/F_\Sigma$	0.326	0.41	0.41	0.39	0.39	0.39	0.27
Bundle length, l, mm	1000	1450	1796	800	800	800	800
p, mm	7.02	7.3	7.68	23.94	23.94	23.94	24.4
$\bar{\mu}$, 1/cm	0.076	0.03	0.034	0.027	0.017	0.0133	0.0063
K	0.025	0.0102	0.0102	0.036	0.023	0.018	0.009

Table 4.1 continued

Parameter	Bundle number						
	13	14	15	16	17	18	19
Re	$2 \cdot 10^3\ 1.2 \cdot 10^4$	$1 \cdot 10^4\ 3 \cdot 10^4$	$4 \cdot 10^3\ 1.5 \cdot 10^4$	$2.5 \cdot 10^4\ 5.2 \cdot 10^4$	$4.2 \cdot 10^4$	$4.2 \cdot 10^4$	$4.2 \cdot 10^4$
Number of rods	37	217	91	7	7	7	7
$x_{St.mix}/d_{rod}$	22.3	16.9	18.5	22.3	21.3	19.7	20.1
l/d_{rod}	126	165	198	29.8	29.8	29.8	31

also comprises the relative lengths of the considered bundles l/d_{rod}. It is seen that $l/d_{rod} \gg x_{st.mix.}/d_{rod}$ for all the bundles. Hence, it may be assumed that the coefficient K will depend mainly on the Fr_M number and the bundle porosity m with respect to heat carrier, i.e.

$$K = K(Fr_M, m)$$

Then the experimental data listed in Table 4.1 may be generalized by the critical relation

$$K = 1.902 Fr_M^{-0.53} m^{1.086} \qquad (4.15)$$

In Fig. 4.1a, a scatter of the experimental data on the porosity m is observed on the coordinates $K = f(Fr_M)$ while in Fig. 4.1b, on the Fr_M number, i.e. the coefficient K increases with m and decreases with increasing the Fr_M number. If the experimental data are compared with (4.15) in the form of $K^* = Km^{-1.086} = f(Fr_M)$, then, as it follows from the inspection of Fig. 4.2, all the experimental points are well in line with one another, which represents formula (4.15).

Thus, formula (4.15) enables one, over the range of $Fr_M = 43-3300$ and after $m \geqslant 0.27$, to calculate the coefficient K necessary to close the system of equations (1.18)–(1.11) when the homogenized flow model is used to determine temperature fields of heat carrier in finned rod bundles in a nonuniform energy release field in the bundle cross-section.

Relation (4.15) is also well consistent with the experimental data on the experimental coefficients K for oval helical tube bundles at $Re \geqslant 10^4$ [39, 9, 16] (Figs. 4.1 and 4.2). Table 4.2 comprises the main similarity numbers and values of the coefficient K [39, 9, 16]. Hence, common criterial relation (4.15) has enabled one to describe an interchannel mixing process both in

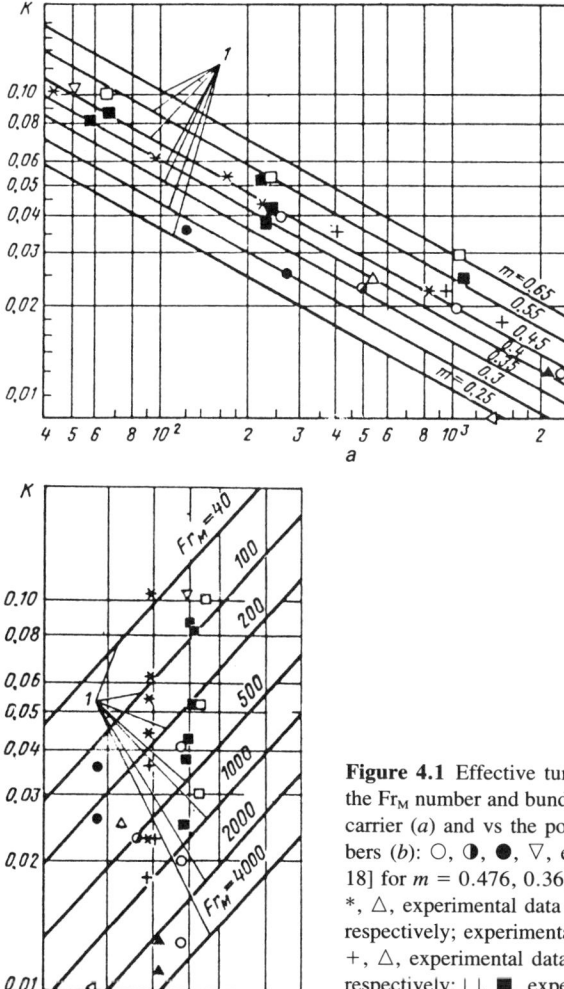

Figure 4.1 Effective turbulent diffusion coefficient vs the Fr_M number and bundle porosity with respect to heat carrier (*a*) and vs the porosity *m* at different Fr_M numbers (*b*): ○, ◐, ●, ▽, experimental data [48, 49, 17, 18] for $m = 0.476$, 0.36, 0.28 and 0.494, respectively; *, △, experimental data [42] for $m = 0.39$ and 0.326, respectively; experimental data [50, 56] for $m = 0.41$; +, △, experimental data [56] for $m = 0.39$ and 0.27, respectively; ⊔, ■, experimental data for $m = 0.527$–0.544 [39, 9, 16] and for $m = 0.477$–0.496 [39], respectively; *1*) relation (4.15).

finned rod and in oval helical tube bundles on the section, on which the coefficient *K* does not depend on the Reynolds number (Re $\geq 10^4$). This points to the same mechanism of heat and mass transfer processes in such bundles.

The experimental data on the coefficient *K* for helical tube bundles are obtained by the method of heat diffusion from a system of linear sources based on the Euler representation of turbulent flow [39, 9, 16]. According to this method, in the case, when a heat release field is nonuniform in the outlet cross-section of a bundle, the temperature fields found from experi-

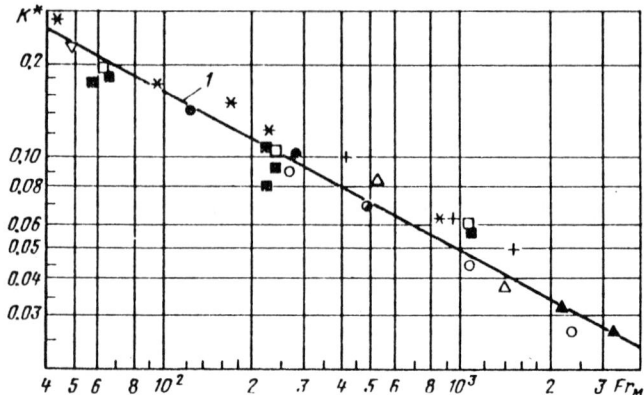

Figure 4.2 Comparison of the criterial relation for the effective turbulent diffusion coefficient with the experimental data for finned rod and helical tube bundles (the notations are the same as in Fig. 4.1).

ment and predicted by solving the system of equations (1.8)–(1.11) under boundary conditions (1.12)–(1.14) are compared, and the most confidence values of the coefficient K_E are determined. In addition, at $Re < 10^4$ the relation

$$K_E = 6.015 Fr_M^{-0.53} m^{1.086} Re^{-0.125} \qquad (4.16)$$

may be recommended. This relation in Fig. 4.3 is compared, at $Re = 8 \cdot 10^3$ and $m = 0.475$, with the expression:

$$K_L = 0.0356 (1 + 8.1 Fr_M^{-0.278}) \qquad (4.17)$$

obtained by the heat diffusion method based on the statistical Lagrange representation of a turbulent field when applied to study a history of individual particles continuously emitted by a point source [39]. It is seen that if at the small Fr_M numbers K_E slightly differs from K_L, then with increasing Fr_M a difference between these coefficients grows. If in accord with [13] an assumption is made that

Table 4.2 Experimental values of the mixing coefficients in helical tube bundles (bundle nos. 1–3 [9, 16], nos. 4–8 [39], heat carrier—air)

Parameter	Bundle number								
	1	2	3	4	5	6	7	8	9
Bundle porosity	0.544	0.539	0.527	0.477	0.492	0.496	0.51	0.51	0.49
Fr_M	63.5	232	1050	1080	236	65	57	222	222
K	0.10	0.053	0.03	0.025	0.043	0.087	0.083	0.053	0.037

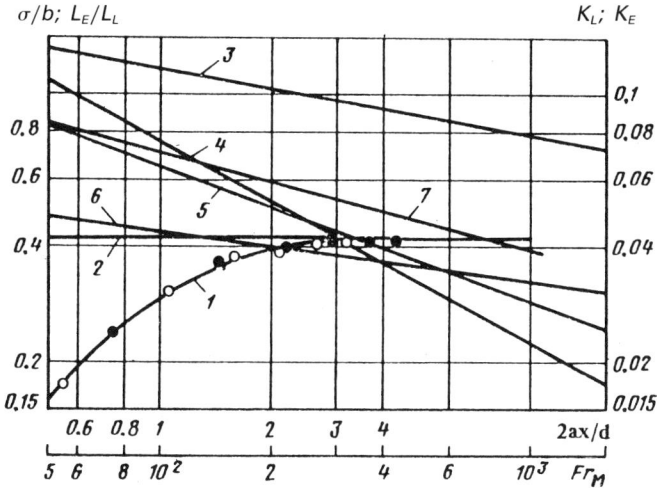

Figure 4.3 Effect of the Fr_M number on heat and mass transfer characteristics in helical tube bundles; *1*) plot of a mean-statistical deviation of temperature distributions as a function of the transformed longitudinal coordinate generalizing the experimental data [58] (●, ○, experimental data [58] for $Fr_M = 314$ and $Fr_M = 1530$; *2–7*, relations (4.24), (4.17), (4.16), (4.19), (4.20) and (4.25), respectively.

$$\frac{K_E}{K_L} = \frac{L_E}{L_L} \tag{4.18}$$

then over the range of the Fr_M numbers ($Fr_M = 50-1500$) we obtain a relationship

$$\frac{L_E}{L_L} = 3.338 Fr_M^{-0.356} \tag{4.19}$$

for spatial turbulence scale ratios. Earlier, in [13], the similarity equation was obtained for the coefficient K_E, (4.12), which has proved to be valid for a more narrow range of $Fr_M = 55-1060$, as compared to (4.16). Using (4.12) and (4.17) arrives at

$$\frac{L_E}{L_L} = 0.785 Fr_M^{-0.127} \tag{4.20}$$

As seen from (4.19) and (4.20), with decreasing the Fr_M number the ratio L_E/L_L increases (Fig. 4.3) since in this case the intensity of turbulent fluctuations [12] grows. The obtained result is identical to the data of Mickelsen who established that the ratio L_E/L_L on the round tube axis increases with a pulsational velocity.

In Fig. 4.3, a noticeable difference between K_L and K_E observed at large Fr_M numbers is probably bound up with use of the method of heat diffusion from a point heat source for a helical tube bundle, with a source size being finite [39]. For a finite-size heat diffusion source the temperature distributions at different distances from it are of the form

$$\frac{T - T_0}{T_{max} - T_0} = \exp\left(\frac{\bar{r}_0^2 - \bar{r}^2}{8\bar{y}^2}\right) f(\bar{r}, \bar{r}_0, \bar{y}^2) \qquad (4.21)$$

$$f(\bar{r}, \bar{r}_0, \bar{y}^2) = \frac{\bar{r}^6 \bar{r}_0^6 + 576\,(\bar{y}^2)^2\,\bar{r}^4 \bar{r}_0^4 + 147.5 \cdot 10^3\,(\bar{y}^2)^4 \bar{r}^2 \bar{r}_0^2 +}{\bar{r}^{12} + 576\,(\bar{y}^2)^2 \bar{r}_0^8 + 147.5 \cdot 10^3\,(\bar{y}^2)^4 \bar{r}_0^4 +} \cdots$$

$$\leftarrow \cdots \frac{+ 9.47 \cdot 10^6\,(\bar{y}^2)^6}{+ 9.47 \cdot 10^6\,(\bar{y}^2)^6} \qquad (4.22)$$

$$\bar{r} = \frac{2r}{b},\ \bar{r}_0 = \frac{2r_0}{b},\ \bar{y}^2 = \frac{y^2}{b^2}$$

different from the Gauss one (at $r_0 = 0$)

$$\frac{T - T_0}{T_{max} - T_0} = \exp\left(-\frac{r^2}{2\bar{y}^2}\right) \qquad (4.23)$$

A dimensionless rms deviation σ/b of distribution (4.21) increases along the jet flowing out of the source (Fig. 4.3), and only at distances of $2ax/d \geq 2.56$ it asymptotically tends to σ/b for distribution (4.23) which is equal to:

$$\sigma/b = 0.423 = \text{const}\,(2ax/d) \qquad (4.24)$$

In (4.24), a is the jet structure coefficient experimentally found in [39] and depending on the Fr_M number according to (4.14). It is seen that the greater the Fr_M number, the larger length of a bundle is needed for a finite-size source be taken as the point one in K_L determination. For the same length of a helical tube bundle with different Fr_M numbers, an accuracy in determining K_L at large Fr_M number will be less. In this case, the methods used in [39] will yield some overestimated value of the coefficient K_L. On the other side, the data on the coefficient K_E for helical tube bundles calculated according to (4.15) and (4.16) are approximately smaller by a factor of 1.5 than those for $Fr_M = 1050$ [9] although these are over the confidence range of the experimental coefficient K_E. Therefore, if the experimental data on coefficient K_E [9] where for $Fr_M = 64$ $K_E = 0.10$, for $Fr_M = 232$ $K_E = 0.053$ and for $Fr_M = 1050$ $K_E = 0.030$ make the basis for comparing K_E and K_L, then the ratios L_E/L_L will be equal to 0.787, 0.535 and 0.387, respectively, and the interpolation formula for L_E/L_L will be of the form (Fig. 4.3):

$$\frac{L_E}{L_L} = 2.263\,Fr_M^{-0.254} \qquad (4.25)$$

In this case, over the range of $Fr_M = 64$–1050 a mean value of $L_E/L_L \approx 0.6$ is set as in the experiments of Mickelsen who found that on the round tube axis over the range Re = $2 \cdot 10^5$–$6 \cdot 10^5$ a mean value of $L_E/L_L \approx 0.6$. This may be expected since in helical tube bundles over the above range of Fr_M numbers at Re $\approx 10^4$, velocity profiles in a wall layer are described by the same power laws as the ones in a round tube at Re $\geqslant 10^5$ [39]. Thus, the experimental data on the coefficient K_E [9] or relations (4.15)–(4.16) which govern the experimental data of various authors with a confidence probability of 0.95 may be employed to calculate temperature and velocity fields.

The suitability of the system of differential equations (1.8)–(1.11) when applied for prediction of the mutual effects of temperatures and velocity fields with a nonuniform heat release field over the bundle radius was proved experimentally. Experimental study of fields T, u, ρu, ρu^2 in the cross-section of a helical tube bundle was made on the experimental setups described in [39, 9]. The number of helical tubes in a bundle was 37 and 127 at heating of a central group consisting of 7 and 37 tubes, respectively. A maximum size of the tube oval was $d = 12.3$ mm. A relative helical tube twisting pitch S/d varied from 6.08 to 26. Experiments were made at Re $\approx 9 \cdot 10^3$–$1.6 \cdot 10^4$, which is characteristic of the considered heat exchangers.

A flow temperature and velocity were measured by a chromel-alumel thermocouple and a Pitot tube mounted on a traverse gear. These fields were measured in the outlet cross-section of a bundle. A velocity field at the bundle inlet was uniform.

Experimental dimensionless velocity and temperature fields are shown in Fig. 4.4a,b. It is seen that a variation character of these parameters is identical but with decreasing the pitch S/d or the Fr_M number, temperature nonuniformities are levelled at higher rates. The fields ρu obtained in terms of the measured values of p, u and T from the state equation are shown in Fig. 4.4c. It is seen that this parameter also varies over the bundle radius. Figure 4.5 presents the experimental fields u and T as well as ρu and ρu^2 for a helical tube bundle with $Fr_M = 232$ for the flow core. Here, comparison is made with the predictions of the system of equations (1.8)–(1.11) done by the method [9]. It is seen that a good agreement between the experimental and predicted fields u, T, ρu and ρu^2 is observed in the flow region where the wall of a helical tube does not exert an influence. Hence, when only a nonuniform temperature field initiated by the nonuniform heat release one is responsible for velocity field nonuniformities in the flow core, the velocity inconsiderably varies over the bundle radius (Fig. 4.5a). At the same time the nonuniformities T, ρu and ρu^2 in the cross-section of a bundle are pronounced (Fig. 4.5a,b,c). Therefore, in calculating temperature and velocity fields in a helical tube bundle by using the homogenized flow model, the axisymmetric problem should be solved by adopting the system of equations (1.8)–(1.11) which describes cylindrical channel flow with volumetric heat release and hydraulic resistance sources, assuming that a velocity vector is parallel to the channel axis. That the boundary condition $\partial u/\partial r|_{r=r_{sh}} = 0$ be satisfied in the framework of the accepted model, the displacement thickness

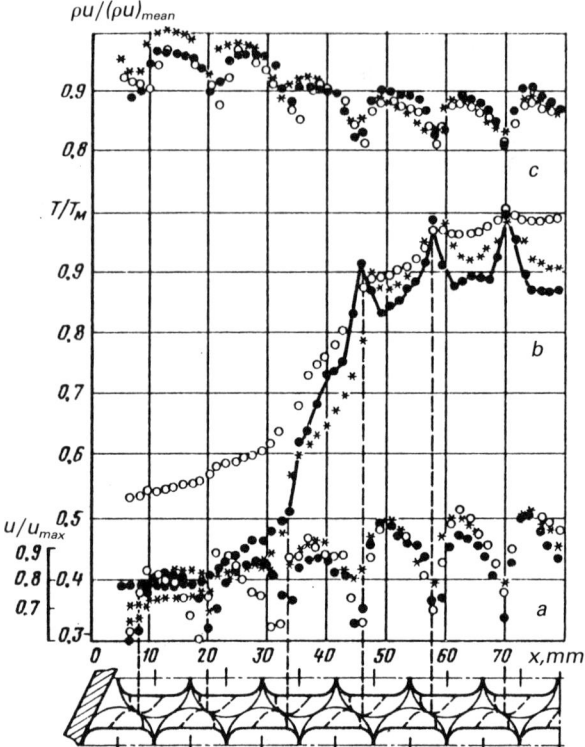

Figure 4.4 Initial fields of a velocity (*a*), temperature (*b*) and a mass velocity (*c*) with $N = 127$ in a helical tube bundle at different Fr_M numbers measured along the oriented rows of tubes: ●, $S/d = 26$, $Re = 1.42 \cdot 10^4$, $q = 2.9 \cdot 10^4$ W/m², *, $S/d = 12.3$, $Re = 9.8 \cdot 10^3$, $q = 2.48 \cdot 10^4$ W/m², ○, $S/d = 6.08$, $Re = 9.74 \cdot 10^3$, $q = 2.53 \cdot 10^4$ W/m².

of a boundary layer must allow for a velocity variation in the wall layer. A material layer of such a thickness is conventionally built up on the tube walls, and homogeneized slip flow is considered under changed boundary conditions. These specific features of the flow model were taken into account in constructing the plots in Fig. 4.5, giving flow core velocities alone.

Figure 4.5a also compares the experimental data obtained on the experimental set-ups with different number of helical tubes in a bundle ($N = 37$ and 127). It is seen that the variation character of u and T for these two bundles is identical, and a scatter of the experimental points is within the predictions at $K = 0.03$–0.09, i.e. a mean value of the coefficient K is almost the same. It should be noted that the system of equations (1.8)–(1.11) reduces if in (1.8) neglect is made of a term responsible for the levelling of velocity field nonuniformities due to turbulent diffusion and due to other transport mechanisms which are allowed for by the coefficient D_t [39] as well as by a term for volumetric hydraulic resistance sources. Then instead of (1.8), we have

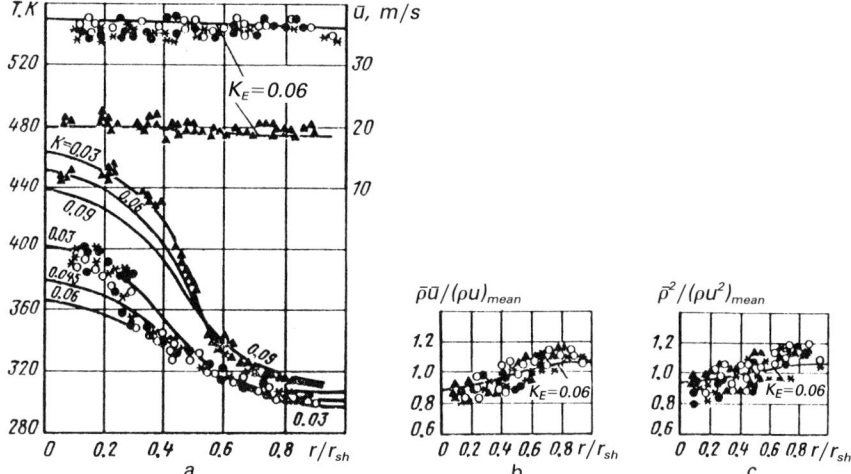

Figure 4.5 Fields of velocity and temperature (a), mass velocity (b) and velocity head (c) in a helical tube bundle at $Fr_M = 232$: ●, ○, *, at $Re = 1.6 \cdot 10^4$, $N = 37$; ▲, $Re \approx 10^4$, $N = 127$; ———, calculation.

$$\frac{dp}{dx} = -\rho u \frac{\partial u}{\partial x} \qquad (4.26)$$

Solving (4.26) for an elementary annular stream and two cross-sections x_1 and x_2, assuming that a density varies with a pressure and a temperature, we have:

$$\frac{\rho_1 u_1^2}{2} + p_1 = \frac{\rho_2 u_2^2}{2} + p_2 - \int_{x_1}^{x_2} \frac{u^2 d\rho}{2} \qquad (4.27)$$

For a section dx long, where $d\rho \approx 0$

$$\frac{\rho u^2}{2} + p = \text{const} \qquad (4.28)$$

Since in the bundle cross-section $\rho = \text{const}$, for two streams i and j we have

$$\rho_i u_i^2 = \rho_j u_j^2 = \rho u^2 = \text{const} \qquad (4.29)$$

Using a relationship between ρ and u in the form of (4.29), the problem may be reduced to solving one energy equation (1.11), which decreases a computer time needed to calculate temperature and velocity fields in a helical tube bundle. However, a choice of a system of equations may be caused only by an agreement between the predicted and experimental data on temperature and velocity fields, mass velocity $(\rho u)_{\text{mean}} = G/F_f$ and velocity head ρu^2, and condition (4.29) is not supported by experiment (Fig. 4.5c). The flow models based on relationship (4.29) are, therefore, inapplicable for calculation of heat and mass transfer in helical tube bundles. At the same time a good coincidence between the experimental velocity and temperature fields, mass velocity and velocity head and those predicted by numerical solution of the system of differential equations (1.8)–(1.11) for flow of a homogenized medium emphasizes the fact that this model, its mathematical description and its calculation method can be employed to determine velocity and temperature distributions in helical tube bundles.

The revealed effect of a heat carrier temperature field initiated by a nonuniform heat-releasing field over the radius of a helical tube bundle on a flow velocity field must be allowed for in developing a flow model and its mathematical description when applied to unsteady heat and mass transfer processes. A necessity to use the motion equation in form (1.8) may be also substantiated in studying the levelling of the velocity field nonuniformity developed by an inlet connecting pipe in the adiabatic air flow conditions. Experiments were made on the models for a heat exchanger consisting of 127 oval helical tubes with the relative pitch $S/d = 16$ and with $Fr_M = 470$ on the experimental set-up detailed in [39]. The flow at the bundle inlet was axisymmetric. Velocity field nonuniformity was promoted by a system of the inlet grids, behind which a turbulence level was 6%. A flow velocity in the outlet cross-sections of the bundles was measured by the total Pitot different-length tube little sensitive to an angle of flow wash up to $\pm 20°$ [39]. The bundle length corresponded to the distances of $11d$, $18.7d$ and $90.5d$ from the inlet. In addition, the inlet conditions remained invariable, $Re \approx 10^4$ and $T_{in} = 305$ K. The rms error in determining a flow velocity was 3%.

It has appeared that the flow core velocity profile at a distance of $11d$ can be described by the 1st kind zero-order Bessel function

$$u_{in}/u_{mean} = 0.8 + 0.35 J_0 (2.405 \bar{r}) \tag{4.30}$$

where $\bar{r} = r/r_{sh}$. In making calculations, this cross-section of a bundle was taken as the initial one. Calculations were made by solving the system of equations (1.8)–(1.11) with boundary conditions (1.12)–(1.14) by the network method using an explicit scheme [9]. Calculated velocity fields over the length and the radius of a bundle are presented in Figs. 4.6 and 4.7. Figure 4.6 compares the curves for a velocity distribution along the bundle at $\bar{r} = 0$ and $\bar{r} = 1$ at different $K \approx 0$, 0.09 and 0.625. It is seen that for $K \approx 0$ the levelling of initial velocity nonuniformity ceases at $x \approx 70d$, for $K_E = 0.09$ at $x = 50d$ and for $K = 0.625$ at $x \approx 32d$, which supports a necessity to use the motion equation in form (1.8).

Figure 4.7 presents the calculations at $K = 0.03$ well consistent with the experimental data for bundle cross-sections $x = 77d$ and $79.5d$ represented as most probable values of a velocity for each r obtained by statistical processing of a large body of experimental data.

It should be noted that the design curves $u = u(r)$ at different K have

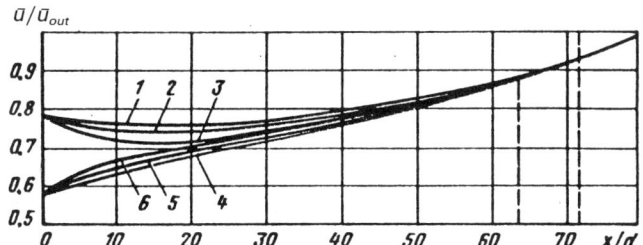

Figure 4.6 Velocity variation along a helical tube bundle: 1–3) calculation at $\bar{r} = 0$ and $K \approx 0.09$ and 0.625, respectively; 4–6, the same at $\bar{r} = 1$.

a small scatter, especially within $K \leq 0.09$, which points to the fact that a value of this coefficient cannot be determined by comparing the experimental and predicted velocity fields, as done in comparing the temperature fields when a great scatter of the design curves (Fig. 4.5) is observed.

In the case of an unsteady process, a contribution of the diffusional term in equation (1.8) may be most pronounced. So, according to [27], when heat power sharply increases at a constant flow rate, the unsteady heat and mass transfer coefficient K_{uns} at the first time moments may, several times, exceed the quasi-stationary coefficient K_{qs} determined by relations (4.15) and (4.16) following the formula

$$\kappa = \frac{K_{uns}}{K_{qs}} = 0.81 \cdot 10^{-4} \text{Fo}_b^{-2} - 0.978 \cdot 10^{-2} \text{Fo}_b^{-1} + 1.21 \quad (4.31)$$

where

$$\text{Fo}_b = \frac{\lambda_b \tau}{c_p \rho_b d_{sh}^2} \quad (4.32)$$

A contribution of the distributed hydraulic resistance $\xi \mu u^2 / 2 d_{eq}$ to levelling velocity nonuniformities may be estimated in the following fashion. Assume that this process is mainly affected by the diffusion term in equation (1.8). Then solving the equation

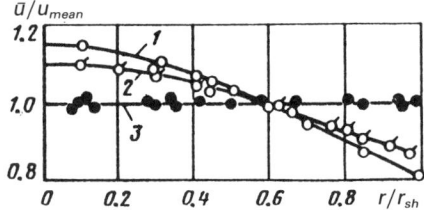

Figure 4.7 Comparison of the experimental and predicted velocities at $K = 0.03$: 1) velocity profile (4.30) at $x = 0$; 2, 3) calculation for $x/d = 7.7$ and 79.5; ○, ○, ●, experimental data for $x/d = 0, 7.7, 79.5$, respectively.

$$\frac{\partial \Delta u}{\partial x} = \frac{D_t}{u}\left(\frac{\partial^2 \Delta u}{\partial r^2} + \frac{1}{r}\frac{\partial \Delta u}{\partial r}\right) \tag{4.33}$$

with boundary conditions

$$\Delta u(0, r) = \Delta u_{max} J_0(2.405\bar{r}) = f(\bar{r}) \tag{4.34}$$

$$\Delta u(x, r)|_{r = r_{sh}} = 0 \tag{4.35}$$

$$\Delta u(x, r) \text{ is finite} \tag{4.36}$$

by the separation-of-variables method arrives at a general solution

$$\Delta u(x, r) = \sum_{n=1}^{\infty} A_n J_0(\mu_n, \bar{r}) \exp\left[-\frac{D_t x}{u}\left(\frac{\mu_n}{r_{sh}}\right)^2\right] \tag{4.37}$$

where

$$A_n = \frac{\int_0^{r_{sh}} f(\bar{r}) J_0(\mu_n \bar{r}) r dr}{\int_0^{r_{sh}} J_0^2(\mu_n \bar{r}) r dr}$$

With regard to (4.34), we obtain $A_1 = 0.35$, $A_2 = A_3 = \ldots A_n = 0$. Finally, a general solution to equation (4.37) yields

$$\Delta u(x, r)/u_{mean} = 0.35 J_0(2.405\bar{r}) \exp\left[-\frac{D_t}{u_{mean}}\left(\frac{2.405}{r_{sh}}\right)^2 x\right] \tag{4.38}$$

Comparing the predicted and experimental velocity distributions arrives at $K = 0.625$. However, in reality, for this bundle, according to (4.15), $K = 0.03$. Hence, a term allowing for the levelling impact of hydraulic resistance in equation (1.8) is important.

The above study has enabled one to substantiate use of the system of differential equations (1.8)–(1.11) and to give recommendations on velocity and temperature field calculations in bundles of helical tubes and spiral-finned rods at a prescribed inlet velocity profile initiated by the inlet connecting pipe.

4.2 HEAT AND MASS TRANSFER IN TWISTED BUNDLES OF HELICAL TUBES

As already mentioned, a heat exchange apparatus with a helical-tube twisted bundle offers providing a more uniform temperature field in the bundle cross-section at azimuthal heat supply nonuniformity due to an additional transfer mechanism responsible for heat carrier flow swirling relative to the bundle axis as against the one with a straight bundle of helical tubes. In this case, heat transfer in a bundle enhances, and hydraulic losses somewhat increase in the intertube space of a heat exchanger. Intensive levelling of temperature field nonuniformities in the bundle cross-section embodies operational reliability of a heat exchanger, and heat transfer enhancement improves its overall dimensions.

Temperature field calculation in twisted bundles calls for studying a heat and mass transfer process and for determining the effective turbulent diffusion coefficient D_t or the dimensionless coefficient K_E found by (4.3) and used to close a system of differential equations for bundle flow.

The method of heat diffusion from a point source described in Section 2.1 was employed to determine the coefficient K_E. This method was also adopted to find K_L for a straight helical tube bundle [39]. In this case, when a turbulent field is given in Lagrange's statistical representation, the mean-statistical squared displacement, $\bar{y}^2$, of heated particles continuously emitted by a diffusion source is determined by the formula

$$\bar{y}^2 = 2\frac{D_t}{u}(x - x_0) \tag{4.39}$$

which is a limiting solution to Taylor's equation for homogeneous and isotropic turbulence at large diffusion times (at large distances from a diffusion source)

$$\frac{d}{dt}\left(\frac{1}{2}\bar{y}^2\right) = \int_{t_0}^{t} \overline{v_1(t_0)v_1(t')}\, dt' \tag{4.40}$$

Application of this method for twisted bundles is justified since the experimental temperature fields in the bundle cross-sections at different distances from a diffusion source are close to normal distribution law (4.23).

Study of heat and mass transfer in a twisted bundle was made on the models of heat exchangers with different twisting principles over the bundle radius: at twisting of all the helical tubes relative to the bundle axis with a constant angle

$$\gamma = \text{const}(r) = \frac{1}{6}\pi \tag{4.41}$$

and at twisting with a constant pitch (m)

$$S_E = \text{const}(r) = 0.65 \tag{4.42}$$

Condition (4.41) results in flow swirling according to $v_\tau = \text{const}(r)$, and condition (4.42), in quasi-solid rotation of hear carrier $v_\tau/r = \text{const}(r)$. In this case, the behaviour of the heat and mass transfer process in bundles with the same dimensions of helical tubes ($S = 171$ mm, $d = 12.2$ mm, $S/d = 14$) which are analyzed in the present section may be different.

A distinctive feature of the investigated bundles also consists in the fact that a twisted bundle according to (4.41) has 61 tubes, $m = 0.62$, $d_{eq} = 12.55$ mm, $Fr_M = 190$ while the one according to (4.42) has 59 tubes, $m = 0.63$, $d_{eq} = 14.19$ mm, $Fr_M = 170$.

The experimental technique was as follows. Heated air was continuously injected via a 17 mm radius round tube (diffusion source) into the cocurrent longitudinal air flow past a tube bundle about 1.5 m long. Hot air was injected at the three radical orientations of a diffusion source: $r_{d.s} = 0$, $r_{d.s} = 19.7$ mm and $r_{d.s} = 38.7$ mm, i.e., at the bundle centre, at the intermediate position and at the bundle periphery. Air temperatures in the bundle cross-sections at the distances of 375, 750 and 1126 mm from a diffusion source were measured by chromel-alumel thermocouples welded to the walls of all the bundle tubes on the inside (Fig. 4.8). Measurement of a tube temperature instead of air one in the flow core has enabled one to refuse from the method of a moving diffusion source [39]. However, as a result there has appeared a systematic error associated with a difference in temperature distributions of heat carrier in the flow core and on the tube walls. This may be illustrated by way of example of a straight helical tube bundle [39], for which dimensionless excess temperature fields in the cross-section at a distance of 0.9 m from a diffusion source for the flow core and according to the measured tube wall temperatures substantially differ. The determined quantities $\bar{y}^2$ and K_L for measured temperatures in the flow core and on the tube walls (Fig. 4.9) evidence that for a straight helical tube bundle with $Fr_M = 314$ the coefficient K_L for a measured flow core temperature is 3.07 times as many as the coefficient K_L calculated in terms of a measured tube wall temperature. This effect must be allowed for in estimating the coefficients K_{twist} for twisted bundles in which tube wall temperatures alone were measured.

The coefficient K_{twist} was measured within $Re = 8.5 \cdot 10^3 - 3.4 \cdot 10^4$ at

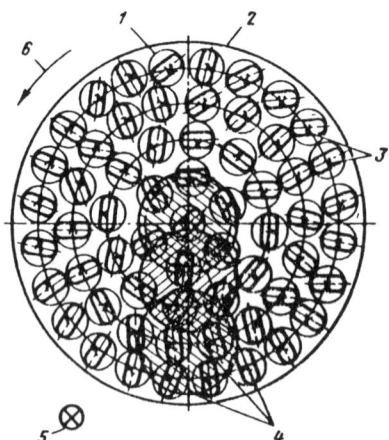

Figure 4.8 Thermocouple location in the cross-section of a twisted bundle of helical tubes at a distance of 0.375 m from the inlet: *1)* helical tube; *2)* shell; *3)* thermocouples welded to the inner tube walls; *4)* possible positions of a diffusion heat source; *5)* heat carrier flow direction in the intertube space of a heat exchanger; *6)* direction of helical tube twisting relative to the bundle axis.

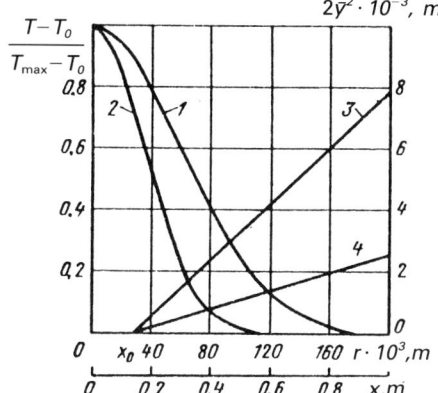

Figure 4.9 Effect of thermocouple locations on temperature distributions in the cross-section of a straight helical tube bundle at $Fr_M = 314$: *1, 2*) dimensionless excess temperature distributions at a distance $x = 0.9$ m from a diffusion source with thermocouples embedded in the flow core and on the tube walls, respectively; *3, 4*) relation (4.39) for temperature distributions 1 and 2, respectively.

air temperatures at the diffusion source outlet $T_{max,x=0} = (394–403)$ K and at the cocurrent flow temperature $T_0 = (294–308)$ K and at a pressure close to atmospheric. The hot air flow rate via the diffusion source amounted to 6% of that of the main flow. So, its contribution to turbulence was neglected.

Experiments were made when the main flow velocity was equal to the one of the jet flowing out of the diffusion source to avoid the impact of the experimental conditions on heat and mass transfer. Moreover, the velocities ranged $u_{d.s} = u_0 = 6.3$–26 m/s, being consistent with the Re number range.

In processing the experimental data it has been allowed for the fact that the nonisothermal jets injected into the bundle on the intermediate and periphery radii are curvi-linear, i.e. the jet axis with a maximum temperature is twisted relative to the bundle axis at an angle of twisting the helical tubes located on these radii. Therefore, the experimental coefficient K_{twist} is determined without regard to ordered transfer by a twisted bundle (with respect to bundle azimuth). This offers comparison of the coefficients K_{twist} with the coefficients K for the bundles of straight helical tubes.

The typical experimental dimensionless excess temperature distributions in the cross-sections of a helical tube bundle twisted according to (4.41) at different distances from a diffusion source on the bundle axis are plotted in Fig. 4.10 where these are compared with the Gauss distribution in the form:

$$\frac{T - T_0}{T_{max} - T_0} = \exp\left[-0.698\left(\frac{r}{r_{mean}}\right)^2\right] \quad (4.43)$$

Coincidence is good, thus pointing to a possible use of the method of heat diffusion from a point source to determine coefficients K_{twist} over the whole range of the main parameters covered by experiment. In plotting the relations $(T - T_0)/(T_{max} - T_0) = f(r/r_{mean})$ and in determining $\bar{y}^2$ and K_{twist}, the

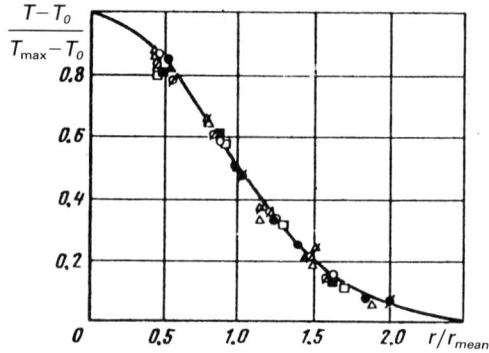

Figure 4.10 Dimensionless excess temperature distributions in the cross-sections of a twisted bundle at $\gamma = \mathrm{const}(r)$ at the axial bundle location of a diffusion source: ●, ○, △, experimental data for $Re = 8.5 \cdot 10^3$; ⬧, ⌀, ▲, the same for $Re = 3.4 \cdot 10^4$; ●, △, ⬧, the same for $x = 0.375$ m; ○, □, ⌀, the same for $x = 0.750$ m; △, □, ▲, the same for $x = 1.126$ m; ———, Gauss' distribution.

experimental data are treated as follows. An arithmetic mean of a temperature has been determined on the radii r_i counted off from the jet axis

$$T_i = \sum_{j=1}^{n} T_j / n \qquad (4.44)$$

where T_j is the temperature measured on the radii r_i at the point j; n is the number of experimental points on the radius r_i. An excess temperature

$$T_i - T_0 = (T_{max} - T_0) \exp(-\beta r_i^2) \qquad (4.45)$$

is measured, where $T_{max} - T_0$ is the excess temperature at $r_i = 0$. Deviations of the experimental points from (4.45) are estimated by the least square method

$$\sum_{i=1}^{n} [(T_i - T_0) - (T_{max} - T_0) \exp(-\beta r_i^2)]^2 = F \qquad (4.46)$$

The quantity F must be minimum, therefore, $\partial F / \partial \beta = 0$ and $\partial F / \partial = (T_{max} - T_0) = 0$. In this case, we have

$$\sum_{i=1}^{n} (-2)[(T_i - T_0) - (T_{max} - T_0) \exp(-\beta r_i^2)](T_{max} - T_0) r_i^2 \exp(-\beta r_i^2) = 0 \qquad (4.47)$$

$$\sum_{i=1}^{n} 2[(T_i - T_0) - (T_{max} - T_0) \exp(-\beta r_i^2)] \exp(-\beta r_i^2) = 0 \qquad (4.48)$$

After the excess temperature $(T_{max} - T_0)$ is found from expression (4.48) and is substituted into (4.47), we obtain an equation to determine β graphically. Then a value of the excess temperature corresponding to the middle jet radius r_{mean} is found

$$T_{mean} - T_0 = (T_{max} - T_0)/2 \qquad (4.49)$$

and from expression (4.45) the quantity

$$r_{mean} = \sqrt{\frac{1}{\beta} \ln \frac{(T_{max} - T_0)}{T_{mean} - T_0)}} \qquad (4.50)$$

is evaluated.

The jet width at the place where $T_i - T_0 = T_{mean} - T_0$ is equal to

$$b = 2r_{mean} \qquad (4.51)$$

For the Gauss distribution

$$\bar{y}^2 = 0.179b^2 \qquad (4.52)$$

Then the coefficient K_{twist} by (4.39) with regard to (4.3) is

$$K_{twist} = \frac{2\bar{y}^2}{4(x - x_0)d_{eq}} \qquad (4.53)$$

In plotting the relations $2\bar{y}^2 = f(x)$, we have

$$2\bar{y}^2 = A + Bx \qquad (4.54)$$

where $A = -4K_{twist} d_{eq} x_0$, $B = 4K_{twist} d_{eq}$, and x_0 is the value of the x-abscissa cut off by straight line (4.54). Experimental data have been processed using the above procedures on the computer 1010 according to the compiled FORTRAN program. Relations (4.54) describing the experimental data are plotted in Fig. 4.11 for a bundle twisted according to (4.41) at different locations of a diffusion source over a bundle radius and at different Reynolds numbers. This figure also compares the experimental data on a dimensionless excess maximum temperature at different distances from a heat diffusion source for a bundle twisted according to (4.41) with the hyperbolic law for decreasing maximum excess temperature along the non-isothermal jet

$$\frac{(T_{max} - T_0)_x}{(T_{max} - T_0)_{x=0}} = \frac{x_0}{x} \qquad (4.55)$$

Good agreement between the experimental data and relation (4.55) (Fig. 4.11) evidences that normal distribution parameters are rather exactly determined using the experimental results obtained by the above procedures since relations (4.43) and (4.55) are the particular solutions to an equation which governs heat diffusion from a point source in the uniform flow. These conditions are implemented for the homogenized flow model [13], considering that for a bundle twisted according to (4.41) longitudinal velocity non-uniformity in the flow core does not exceed ±20% [39] as in the case of a straight helical tube bundle.

Experimental coefficients K_{twist} for a bundle twisted according to (4.41) determined by the cited procedures are presented in Table 4.3.

Inspection of Table 4.3 shows that for this bundle the coefficient K_{twist}

104 UNSTEADY HEAT AND MASS TRANSFER IN HELICAL TUBE BUNDLES

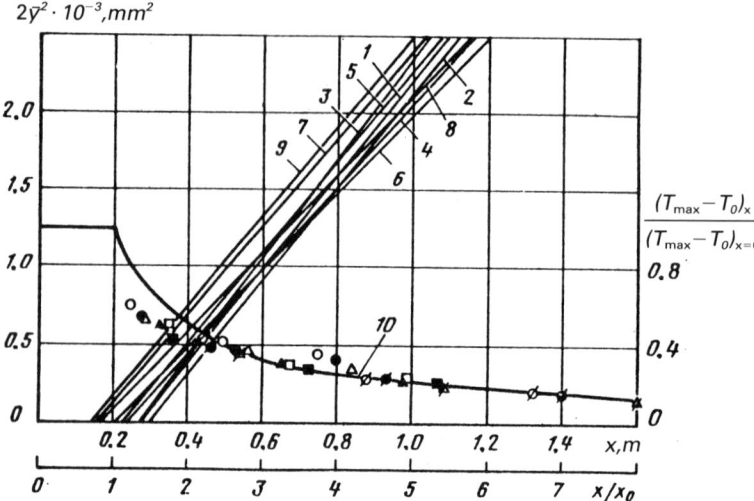

Figure 4.11 Nonisothermal jet length variation of a mean-statistical squared particle path and a dimensionless excess maximum temperature for a bundle with $\gamma = \text{const}(r)$: *1–9*) experimental relations (4.54); *10*) relation (4.55); ●, ○, △, ▲, □, ■, ◆, ⌀, ♠, experimental data for dimensionless excess maximum temperature; *1–3*) ●, ○, △, for diffusion source locations at $r_{d.s}/r_{sh} = 0$; *4–6*) ▲, □, ■, the same at $r_{d.s}/r_{sh} = 0.325$; *7–9*) ◆, ⌀, ♠, the same at $r_{d.s}/r_{sh} = 0.638$; 1, 4, 7, ●, ▲, ◆, at $Re = 8.5 \cdot 10^3$; 2, 5, 8) ○, □, ⌀, at $Re = 1.46 \cdot 10^4$; 3, 6, 9) △, ■, ♠, at $Re = 3.4 \cdot 10^4$.

does not evidently depend on the Re number and the ratio $r_{d.s}/r_{sh}$, and the observed deviations lack any regular pattern. Then it may be, within $\pm 12\%$, considered that in the experimental Re number range the coefficient K_{twist} is equal to an arithmetic mean of the quantities given in Table 4.3, i.e.

$$K_{twist} = 0.0573 = \text{const}(Re, r_{d.s}/r_{sh}) \quad (4.56)$$

Obtained results (4.56) are similar to the data on the effect of the Re number and the ratio $r_{d.s}/r_{sh}$ upon the coefficient K_L in straight helical tube bundles [39] over the investigated Re number range. This may be expected because the presence of through and spiral channels is mainly responsible for longitudinal velocity field nonuniformity in a bundle twisted according to (4.46)

Table 4.3 Experimental values of K_{twist} at $\gamma = \text{const}(r)$

Reynolds number	Position of a diffusion source		
	$r_{d.s}/r_{sh} = 0$	$r_{d.s}/r_{sh} = 0.325$	$r_{d.s}/r_{sh} = 0.638$
$8.5 \cdot 10^3$	0.0620	0.0535	0.0575
$1.46 \cdot 10^4$	0.0608	0.0579	0.0506
$3.4 \cdot 10^4$	0.0644	0.0507	0.0586

which is equal to ±20% as in a bundle of straight helical tubes [39]. Hence, at $\gamma = 20° = \mathrm{const}(r)$ or $S_{\mathrm{twist}}/2r = 8.6$ the additional longitudinal velocity nonuniformity over a bundle radius is not almost observed. As for a numerical value of K_{twist}, the quantity 0.0573 is underestimated because of temperature measurements on the tube wall but not of heat carrier in the flow core (Fig. 4.9). From [39], the coefficient K_L determined in Lagrange's flow representation

$$\dot{K}_L = 0.0356(1 + 8.1\ \mathrm{Fr}_M^{-0.278}) \tag{4.57}$$

is equal to 0.1027 at $\mathrm{Fr}_M = 190$. If a correction similar to the one for $\mathrm{Fr}_M = 314$ and equal to 3.07 (Fig. 4.9) is made for the coefficient $K_{\mathrm{twist}} = 0.0573$, then the coefficient K_{twist} is probably overestimated as a heat carrier temperature difference in the flow core and on the tube wall decreases with Fr_M. At the same time, since the mechanism for ordered transfer with respect to a bundle azimuth is not allowed for in determining the coefficient K_{twist} because a curvilinear jet is considered, the same mechanisms as those in a straight helical tube bundle [13] will be responsible for mixing. It may be, therefore, assumed that $K_{\mathrm{twist}} = K_L$. Then a correction for a temperature distribution difference in the flow core and on the tube wall at $\mathrm{Fr}_M = 190$ will be equal to the ratio $K_L/K_{\mathrm{twist}} = 0.1027/0.0573 = 1.7923$.

A system of differential equations based on Euler's turbulent flow representation [13] is used to calculate temperature fields in helical tube bundles. To close this system of equations, it is necessary to know the coefficients determined by the method [13] since in a general case $K_L \ne K_E$. For a bundle with straight helical tubes the coefficient K_E may be found by formulas (4.15) and (4.16). At $\mathrm{Fr}_M = 190$, $m = 0.62$ and $K_E = 0.0702$, i.e. $K_E/K_L = 0.684$. Assuming that such a ratio also remains for a bundle twisted according to (4.41), we obtain that $K_{\mathrm{twist}} = 0.0702$. This value of the coefficient K_{twist} is mean over the bundle irrespective of a place, at which there appears temperature nonuniformity, and it may be used to calculate temperature fields in bundles twisted according to (4.41). Azimuthal transfer in such a bundle is allowed for by including, into the energy equation, the terms responsible for a temperature variation along the φ-coordinate. This equation is of the form:

$$\rho u c_p \frac{\partial T}{\partial x} + \rho c_p \frac{v_T}{r} \frac{\partial T}{\partial \varphi} = \frac{1}{r}\frac{\partial}{\partial r}\left(\rho c_p D_t r \frac{\partial T}{\partial r}\right)$$
$$+ \frac{1}{r^2}\frac{\partial}{\partial \varphi}\left(\rho c_p D_t \frac{\partial T}{\partial \varphi}\right) + q_v \frac{1-m}{m} \tag{4.58}$$

Accordingly, the motion equations also take into account a parameter variation along the coordinate

$$\rho u \frac{\partial u}{\partial x} + \rho \frac{v_\tau}{r} \frac{\partial u}{\partial \varphi} = -\frac{\partial p}{\partial x} - \frac{\xi \rho u}{2 d_{eq}} \sqrt{u^2 + v_\tau^2}$$

$$\rho u \frac{\partial v_\tau}{\partial x} + \frac{\rho v_\tau}{r} \frac{\partial v_\tau}{\partial \varphi} = -\frac{\xi \varphi}{2 d_{eq}} \rho v_\tau \sqrt{u^2 + v_\tau^2} \qquad (4.59)$$

The experimental values of the coefficients K_{twist} in a bundle twisted according to (4.42) have been also processed by the cited procedures since the measured temperature distributions have obeyed relations (4.43), (4.54) and (4.55). Experimental values of the coefficients K_{twist} for this bundle are compiled in Table 4.4.

From Table 4.4 it is seen that for a bundle with $S_{twist} = \text{const}(r)$ the coefficient K_{twist} vs the Re number and the diffusion source location $r_{d.s.}/r_{sh}$ may be represented as:

$$K_{twist} = 0.6 \, \text{Re}^{-0.243} - 0.0161 r_{d.s.}/r_{sh} \qquad (4.60)$$

With increasing the Re number, the coefficient K_{twist} decreases even in the Re number range, over which there exists the self-similarity of K with respect to the Re numbers [13] for straight helical tube bundles. As far as the diffusion source moves from the bundle centre to its periphery the coefficient K_{twist} decreases, too. Such laws for the coefficient K_{twist} may be attributed to the tube twisting principle, (4.42), affecting heat and mass transfer in a bundle. The thing is that the relative twisting pitch of helical tubes $S_{twist}/2r$ in this case decreases from the bundle centre to its periphery according to the hyperbolic relation

$$S_{twist}/2r = 5.57/(r/r_{sh}) \qquad (4.61)$$

and the angle of twisting of helical tubes relative to the bundle axis, accordingly, increases with a bundle radius. In this case, a longitudinal velocity component varies in correspondence with the curve having a maximum in the peripheral part of a bundle. This is seen from Fig. 4.12, giving a bundle radius variation of an excess longitudinal velocity based on its mean-mass value in a bundle. Such a behaviour of velocity distributions exists behind an axial-blade swirler, near which the specific features of flow were analyzed in [47]. A rotational velocity component in a bundle accord-

Table 4.4 Experimental values of K_{twist} at $S_{twist} = \text{const}(r)$

Reynolds number	Position of a diffusion source		
	$r_{d.s.}/r_{sh} = 0$	$r_{d.s.}/r_{sh} = 0.325$	$r_{d.s.}/r_{sh} = 0.638$
$8.5 \cdot 10^3$	0.0668	0.0629	0.0560
$1.46 \cdot 10^4$	0.0576	0.058	0.0436
$3.4 \cdot 10^4$	0.0440	0.0415	0.0359

ing to flow swirling principle (4.61) should vary practically following the quasi-solid law of rotation. Since swirled flow is peculiar to the fact that a radial velocity component is much less than the components u and v_τ [47], a radial static pressure gradient is determined by the equation

$$\frac{\partial p}{\partial r} = \rho \frac{v_T^2}{r} \qquad (4.62)$$

The value of this gradient may be very substantial. In this case, there appears a pressure drop between the peripheral and axial flow regions. Nonuniform velocity profiles v_τ and u (Fig. 4.12) generate turbulence, whose intensity is different in different flow regions. So, according to [47], near axial-blade swirlers the intensity of velocity pulsations in the peripheral region of a channel is 4–7% while in the axial region it sharply increases up to 30–35%. So, the main reason for increasing K_{twist} in the central region may be attributed to substantial flow turbulization due to flow swirling. In the peripheral bundle region where the turbulence intensity is remarkably lower than the axial one, the coefficient K_{twist} decreases. Thus, the specific features of the swirled flow structure in a bundle according to (4.61) are responsible for K_{twist} as a function of dimensionless radius r/r_{sh}. Similarly, K_{twist} decreases with increasing the Re number since with increasing the Re number, the turbulence intensity decreases simultaneously in all the zones, with this quantity being nonuniform over the bundle radius. It is known that in a straight helical tube bundle the turbulence intensity also drops [12] with a Re number increase, however, this does not alter K at Re $\geqslant 10^4$. This difference is apparently bound up with the simultaneous impact of flow swirling according to (4.61) both on the turbulence intensity and on the turbulence scale which decreases with increasing Re, too.

A correction being introduced for the effect of the location of thermocouples used for measuring the temperature distributions in the bundle cross-section similar to the one for a bundle with $\gamma = \text{const}(r)$, we obtain a formula at $S_{twist} = \text{const}(r)$ in Lagrange's turbulent flow representation

$$K_{twist} = 1.075 \, \text{Re}^{-0.243} - 0.0289 r_{d.s}/r_{sh} \qquad (4.63)$$

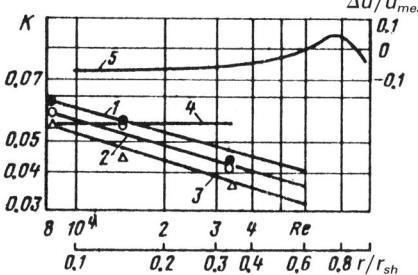

Figure 4.12 Dimensionless effective turbulent diffusion coefficient vs different factors: ●, ○, △, experimental data for a bundle with $S_{twist} = \text{const}(r)$ at $r_{d.s}/r_{sh} = 0$, 0.325 and 0.638, respectively; 1–3) relation (4.60) for diffusion source positions $r_{d.s}/r_{sh} = 0$, 0.325 and 0.638, respectively; 4) relation (4.56) for a bundle with $\gamma = \text{const}(r)$; 5) bundle radius variation of an excess velocity at $S_{twist} = \text{const}(r)$.

Considering that the turbulence scales in Euler's and Lagrange's representations are related by the quantity equal to 0.684, we obtain a relation

$$K_{twist} = 0.735\, Re^{-0.243} - 0.0198 r_{d.s}/r_{sh} \qquad (4.64)$$

for calculation of K_{twist} at $S_{twist} = \text{const}(r)$. Thus, analysis of heat and mass transfer has evidenced that in bundles of helical tubes twisted according to $\gamma = \text{const}(r)$ and $S_{twist} = \text{const}(r)$ there exists a substantial difference in the values of K_{twist} as a function of main parameters (Fig. 4.12) being attributed to different heat carrier flow conditions and, hence, to a difference in the flow structure in these bundles.

The discovered specific features of heat and mass transfer processes in bundles composed of helical tubes differently twisted relative to the bundle axis offer recommendations on practical use of bundles with a constant angle of twisting of helical tubes since in this case more intensive mixing of heat carrier is provided in the bundle cross-section irrespective of the places, at which heat supply nonuniformities develop.

4.3 HEAT TRANSFER AND HYDRAULIC RESISTANCE IN HELICAL TUBE BUNDLES AT UNIFORM HEAT SUPPLY

Heat transfer in helical tube bundles was studied in a number of monographs [10, 39, 40, 51, 52]. In generalizing the experimental heat transfer data, different theoretical considerations checked experimentally were used to derive similarity equations. So, in [10], based on the similarity and dimension theories, assuming that the heat carrier flow is swirled by the spiral tube channels according to the solid-state law $v_\tau/r = \text{const}$, a number is proposed for a relationship between the inertia and centrifugal forces in a bundle of helical tubes:

$$Fr = S^2/(2\pi^2 dd_{eq})$$

which may be, accurate to a constant, represented as

$$Fr_M = S^2/(dd_{eq}) \qquad (4.65)$$

The diversified features of the bundle geometry, (4.65), offer extending the experimental data on heat transfer and hydraulic resistance to the non-similar helical tube bundles [10]. Using the FR_M number, we developed a procedure of calculating heat transfer which is based on an effective wall layer thickness

$$\delta = 0.5(1 + 3.6\, Fr_M^{-0.357})^{-4} d_{eq} \qquad (4.66)$$

taken as a characteristic size with heat carrier flowing in a helical tube bundle. In this case, the data on heat transfer in bundles at $Re_\delta > 5 \cdot 10^2$ and $Fr_M > 90$ are generalized by the relation for round tubes [10]

$$\mathrm{Nu}_\delta = 0.020 \, \mathrm{Re}_\delta^{0.8} \, \mathrm{Pr}^{0.4} \qquad (4.67)$$

plotted in Fig. 4.13. In the transition flow region at $\mathrm{Re}_\delta < 5 \cdot 10^2$ and $\mathrm{Fr}_M \geq 64$ heat transfer is determined by the expression

$$\mathrm{Nu}_\delta = 6.47 \, \mathrm{Fr}_M^{-0.845} \, \mathrm{Re}_\delta^n \, \mathrm{Pr}^{0.4} \qquad (4.68)$$

where

$$n = 0.212 \mathrm{FR}_M^{0.194} \qquad (4.69)$$

at $\mathrm{Fr}_M \geq 924$ $n = 0.8$. With decreasing the Fr_M number a power at Re_δ in (4.69) reduces from 0.8 at $\mathrm{Fr}_M = 924$ to 0.475 at $\mathrm{Fr}_M = 64$. At $\mathrm{Fr}_M < 90$ [10, 39] heat transfer in a bundle enhances additionally (Fig. 4.13), and at $\mathrm{Fr}_M = 64$ and $\mathrm{Re}_\delta > 500$ it is described by the formula

$$\mathrm{Nu}_\delta = 0.0248 \, \mathrm{Re}_\delta^{0.8} \, \mathrm{Pr}^{0.4} \qquad (4.70)$$

In (4.67), (4.68) and (4.70) the Nu_δ and Re_δ numbers are of the form

$$\mathrm{Nu}_\delta = \alpha \delta / \lambda, \quad \mathrm{Re}_\delta = \rho u_{\mathrm{mean}} \delta / \mu$$

The above-mentioned experimental data processing extending the modelling potentialities of heat transfer is supported by the experimental results on the flow structure in helical tube bundles. In [39], it is emphasized that with heat carrier flowing in helical tube bundles, a thin wall layer is formed on the tube walls, and the flow core has approximately a constant velocity. The wall layer thickness δ decreases with decreasing Fr_M. In this case, the wall layer flow is swirled according to the law $v_\tau / r = \mathrm{const}$, and flow core swirling is specified by interacting spiral flows in adjacent tubes [3]. The discovered specific features of flow also support the proposed flow model [51] which is based on the semi-empirical Prandtl turbulence theories and is concerned with the interaction of two flows directed at an angle to one another in a flat equivalent channel. In the middle zone of this channel where a tangential velocity component vector is in a reverse direction, turbulence and secondary flows are generated in addition to the mechanisms responsible for the turbulence onset typical of a straight channel. Moreover, shear stresses are resolved into axial and tangential components [51]:

$$\tau_x = (\mu + \mu_{\mathrm{T}x}) \, du_x / dy, \quad \tau_z = (\mu + \mu_{\mathrm{T}z}) \, du_z / dy \qquad (4.71)$$

assuming that $\mu_{\mathrm{T}x} = \mu_{\mathrm{T}z} = \mu_\mathrm{T}$ and using Yu. A. Koshmarov's hypothesis for the rotational-translational liquid motion between two co-axial cylinders rotating about one another:

$$\mu_\mathrm{T} = \rho l^2 \left\{ |du_x/dy| + |du_z/dy| \right\} \qquad (4.72)$$

Near a wall $l = ky$, in the middle channel zone $l = ky_1 = \mathrm{const}$ where $\bar{y}_1$ corresponds to a point with $u_z = u_{z\,\mathrm{max}}$.

Wall shear stress may be resolved into two components: axial τ_{xw} and

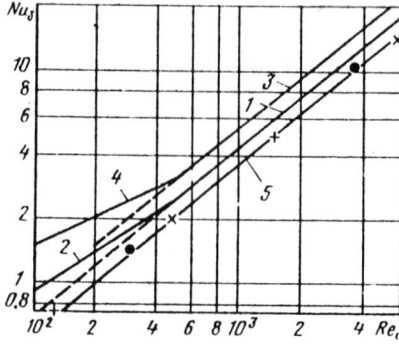

Figure 4.13 Comparison of the experimental data on heat transfer from bundles with different number of tubes using the flow model based on the concept of a characteristic wall layer thickness: *1)* relation (4.67) for $N \geq 37$; 2) relation (4.68) at $Fr_M = 232$; *3, 4)* relations (4.70) and (4.68) at $Fr_M = 64$; *5)* relation (4.83) for a bundle with $N = 19$; $+$, $\bullet$, $\times$, points confining the experimental range of $Re = 5 \cdot 10^3 - 6 \cdot 10^4$ at $Fr_M = 26, 98, 235$ for bundles with $N = 19$.

tangential τ_{zw}. At the channel centre, shear stress consists of the tangential components τ_{z0} alone. As known, the steady stabilized incompressible liquid flow in a flat channel is described by a linear shear stress distribution in height. Similarly, the tangential shear stress component τ_z varies linearly from τ_{zw} on the wall to τ_{z0} on the channel axis. In this case, at the point $y = y_1$ the shear stress $\tau_z = 0$. Then for the stabilized flow we have

$$\tau_x = \tau_{xw}(1 - y/y_0)$$
$$\tau_z = \tau_{zw} + (\tau_{z0} - \tau_{zw})(y/y_0) \tag{4.73}$$

where y_0 is the distance from the channel axis.

Solving the system of equations (4.73) with regard to (4.71) and (4.72) allows axial and tangential velocity component fields at prescribed Re and $\Gamma = u_{z\,mean}/u_{x\,mean} = \pi^2/(S^2/dd_{eq}) = \pi^2/Fr_M$ as well as total hydraulic losses to be found from the force balance equation

$$\Delta p_\Sigma = \Delta p_x + \Delta p_z = [\tau_{xw} + (\tau_{zw} + \tau_{z0}B)\Gamma]\Delta x/y_0 \tag{4.74}$$

where the coefficient B takes into account the real geometry of a bundle cell.

In determining a relation for heat transfer calculations, consideration is made of the energy equation for one-dimensional flow in the direction of the total velocity vector $u_\Sigma = \sqrt{u_x^2 + u_z^2}$:

$$\frac{\partial}{\partial y}\left[(\lambda + \lambda_T)\frac{\partial T}{\partial y}\right] = c\rho u_\Sigma \frac{\partial T}{\partial x} \tag{4.75}$$

In a dimensionless form for $q_w = $ const we obtain

$$Nu = 4 / \int_0^1 \frac{(\int_1^{\bar{\eta}} \bar{u}_\Sigma d\bar{\eta})^2}{1 + (Pr/Pr_T)(\mu_T/\mu)} d\bar{\eta} \tag{4.76}$$

where $\bar{u}_\Sigma = u_\Sigma/u_{\Sigma\,mean}$ is the dimensionless velocity.

Integration of (4.76) is made with respect to layers: 1) $\mu_T/\mu \ll 1$, viscous layer; 2) $\mu_T/\mu \gg 1$, $l = ky$; 3) $\mu_T/\mu \gg 1$, $l = ky_1 = $ const. Calculation results for equations (4.74)–(4.76) are approximated by the approximate relations for ξ and Nu which for 19-tube bundles with $m = 0.45$ and $Fr_M = 26, 98, 235$ ($S/d = 4.15, 8.3, 12.45$) at the stabilized flow length generalize the experimental data [39]:

$$\xi = \frac{0.25}{Re^{0.22}} \left[(1 + \frac{\pi^2}{0.9 Fr_M})^{1.5} + \frac{100}{Fr_M^{1.25}} \right] \psi \qquad (4.77)$$

$$Nu = 0.35 Re^{0.75} \left(1 + \frac{\pi^2}{0.5 Fr_M}\right)^4 \left(1 + \frac{1.3}{Fr_M^{0.6}}\right) \left(\frac{T_w}{T_f}\right)^{-n} \qquad (4.78)$$

where

$Nu = \alpha d_{eq}/\lambda_f$, $Re = \rho u_{mean} d_{eq}/\mu_f$

In (4.77) and (4.78) for $4.15 < S/d < 8.3$

$$\psi = 1 - (1 - \psi_0)(S/d - 4.15)/4.15 \qquad (4.79)$$

$$\psi_0 = (T_w/T_f)^{-(\lg Re - 4)} \qquad (4.80)$$

for $S/d \geq 8.3$ $\psi = \psi_0$ \qquad (4.81)

at $4.15 < S/d < 12.45$

$$n = 0.55 - 0.0663(12.45 - S/d) \qquad (4.82)$$

at $S/d > 12.45$ $n = 0.55$, at $S/d < 4.15$ $n = 0$.

The experimental data for a 19-tube bundle may be also generalized by the formula

$$Nu_\delta = 0.0167 Re_\delta^{0.8} Pr^{0.4} \qquad (4.83)$$

The values of the Nu_δ number calculated by formula (4.83) are less than those obtained by relation (4.67) for bundles with the tube number $N \geq 37$ (Fig. 4.13). Such a difference may be attributed to the effect of a peripheral tube row found by different procedures of experiment and experimental data processing. It is known that the conditions for heat transfer from a peripheral tube row or from rods [57] are similar to those for liquid circular channel flow, with only one wall being heated, when the heat transfer coefficient decreases. Probably, this effect is also observed in 19-helical tube bundles when the heat transfer coefficient is determined in terms of an arithmetic mean wall temperature of all the 19 tubes and in terms of a mean-calorimetric liquid temperature. In bundles with the number of helical tubes equal or greater than 37, the heat transfer coefficient is found through a measured wall temperature of the central tube alone.

A decrease in the heat transfer coefficient on a peripheral row of spiral tube arrangement rods was also revealed in [57] dealing with a 7-rod bundle for $S/d = 12, 24, 36$ ($Fr_M = 220, 860, 1900$), $p/d = 1.237$ and $Pr = 0.7$.

In this case, the heat transfer coefficient was investigated separately for the central and peripheral rods. These data may be compared with those on heat transfer in a helical tube bundle since a flow pattern in bundles of spiral tube arrangement rods and of helical tubes is the same, and heat and mass transfer characteristics are generalized by relation (4.15). To make comparison, let us represent the experimental data for bundles with $N \geq 37$ at the stabilized flow length by the similarity equation [10] as:

$$Nu = 0.023\ Re^{0.8}\ Pr^{0.4}(1 + 3.6\ Fr_M^{-0.357})(T_w/T_f)^{-0.55} \tag{4.84}$$

The Nusselt number ratio of relation (4.84) and the formula for round tubes

$$Nu_{tube} = 0.023\ Re^{0.8}\ Pr^{0.4}(T_w/T_f)^{-0.55} \tag{4.85}$$

yields the expression

$$A_1 = Nu/Nu_{tube} = 1 + 3.6\ Fr_M^{-0.357} \tag{4.86}$$

which is compared with the experimental data [10, 39] in Fig. 4.14a. If formula (4.78) is represented as

$$Nu = 0.044\ Re^{0.75}\left(1 + \frac{\pi^2}{0.5\ Fr_M}\right)^{0.4}\left(1 + \frac{1.3}{Fr_M^{0.6}}\right)\left(\frac{T_w}{T_f}\right)^{-0.55} \tag{4.87}$$

and is divided by (4.85), we obtain the expression

$$A_2 = \frac{1.913}{Pr^{0.4}Re^{0.05}}\left(1 + \frac{\pi^2}{0.5\ Fr_M}\right)^{0.4}\left(1 + \frac{1.3}{Fr_M^{0.6}}\right) \tag{4.88}$$

which at $Re = 10^4$ and $Fr_M = 20-1000$ is well consistent with (4.86). This points to the fact that heat transfer mechanisms are equally taken into account in relations (4.78) and (4.84) but accurate to a constant value. So, the (4.78) and (4.85) ratio gives the relationship:

$$A_3 = \frac{1.52}{Pr^{0.4}Re^{0.05}}\left(1 + \frac{\pi^2}{0.5\ Fr_M}\right)^{0.4}\left(1 + \frac{1.3}{Fr_M^{0.6}}\right) \tag{4.89}$$

which well describes the experimental data obtained for bundles composed of 19 helical tubes (Fig. 4.14a) and lies below relation (4.86) for bundles with $N \geq 37$. This coincides with the plot in Fig. 4.13 where comparison is also made of the data for bundles with $N = 19$ and above or $N = 37$.

Figure 4.14a also plots the experimental data [57] for 7-rod bundles at $Re = 10^4$. It is seen that the results on the Nu number for the central rod with spiral tube arrangement well conform with (4.86) and those on Nu for peripheral rods, with relation (4.89). This evidences the same effect of a peripheral tube row on the heat transfer coefficient in bundles of helical tubes and spiral tube arrangement rods at $Re \approx 10^4$. In this case, heat transfer from bundles with a great number of spirally finned rods may be calculated by formulas (4.67) and (4.84) only at $Re < 3 \cdot 10^4$. Indeed, at Re

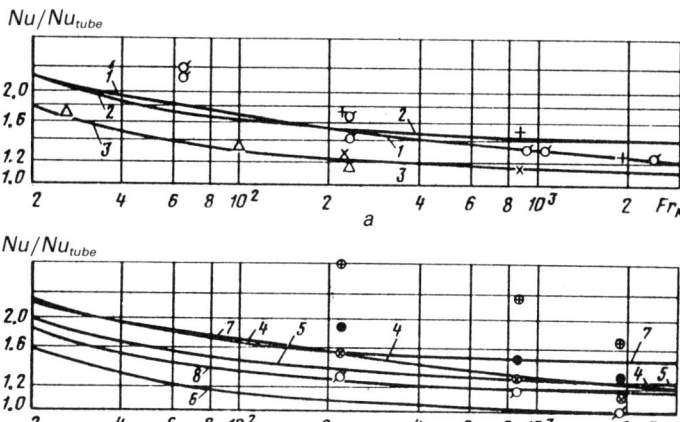

Figure 4.14 Comparison of the experimental data on heat transfer at Re = 10^4 (a) and at Re = $5 \cdot 10^3$ and 10^5 (b) for bundles with different number of tubes using the flow model allowing for a transverse velocity component: 1–3) relations (4.86), (4.88) and (4.89) at Re = 10^4; 4–6) the same at Re = 10^5; 7–8) relations (4.88) and (4.89) at Re = $5 \cdot 10^3$; ϕ, $\triangle$, experimental data [10, 51] for bundles with $N \geq 37$ and 19, respectively; +, ×, experimental data [57] for central and peripheral spirally finned rods at $N = 7$ and Re = 10^4; ⊕, ⊗, the same at Re = 10^5; ●, ϕ, the same at Re = $5 \cdot 10^3$.

> $3 \cdot 10^4$ in [57] a relative increase of the Nu number is observed, as compared to the relation Nu = f(Re) over the range of Re < $3 \cdot 10^4$, which is probably characteristic of heat transfer in bundles of spirally finned rods. This is clearly seen in Fig. 4.14b where comparison is made between the experimental data [57] for Re = $5 \cdot 10^3$ and 10^5 and relations (4.86), (4.88) and (4.89). If for Re = $5 \cdot 10^3$ the results [57] practically coincide with the relations for helical tube bundles (Fig. 4.14b) similarly as at Re = 10^4 (Fig. 4.14a), then for Re = 10^5 the data [57] for a central rod lie much above relations (4.86) and (4.88), and the results for peripheral rods agree with relation (4.86). At the same time relation (4.89) at Re = 10^5 lies much below than at Re = $5 \cdot 10^3$ and 10^4. It is apparent that the extrapolation of the data for 19-tube bundles at Re < $6 \cdot 10^4$ [39, 51] and of the results [10, 39] with $N \geq 37$ at Re < $4 \cdot 10^4$ to higher Re numbers is not warranted or a sharp heat transfer increase at Re > $3 \cdot 10^4$ in bundles of spirally finned rods is associated with the flow features in the investigated bundles of rods.

It should be noted that a difference in the heat transfer data for bundles with $N = 19$ and ≥ 37 helical tubes may be somewhat affected by a periodicity in varying the Nu number along the bundle and at the stabilized flow length as well [39]. This phenomenon is bound up with a spatial flow pattern in a helical tube bundle and with a periodic change in the flow past a spiral surface of tubes along their length. The specific features of heat transfer at the starting flow length in a helical tube bundle with axisymmetric inlet heat carrier flow are considered in [39].

Thus, this analysis has emphasized that heat transfer from a helical tube in a bundle depends on its orientation. Heat transfer from the peripheral tubes being in contact with the shell is somewhat less than the one from the central tubes. A decrease in mean heat transfer from a bundle due to the effect of the peripheral tubes is the less, the greater is the number of tubes in a bundle. Formulas (4.84) or (4.67)–(4.70) are, therefore, recommended to calculate mean heat transfer from a bundle with a great number of tubes ($N \geqslant 37$). If the number of tubes in a bundle is less ($N \leqslant 19$), formula (4.78) is recommended.

Hydraulic resistance in helical tube bundles must not depend on a number of tubes when different data processing procedures are applied. So, in [10, 39], the formulas:

$$\xi = \frac{0.3164}{\mathrm{Re}_d^{0.25}} (1 + 3.6 \mathrm{Fr}_M^{-0.357}) \tag{4.90}$$

$$\xi = 0.266/\mathrm{Re}_\delta^{0.25} \tag{4.91}$$

were recommended to calculation ξ at $\mathrm{Fr}_M > 90$.

If (4.90) is based on ξ_{tube} found by the Blasius formula for round tubes, then expression (4.86) will be obtained which is well consistent with the experimental data [10, 39, 43] at $\mathrm{Fr}_M > 90$ for helical tube bundles with $N = 19$, 37 and 127 at $\mathrm{Re} = 10^4$. The data [51] on ξ (formula (4.77)) based on ξ_{tube} by the Blasius formula at $\mathrm{Re} = 10^4$ and $\mathrm{Fr}_M = 98$ and 235 are rather well governed by formula (4.86). However, at $\mathrm{Fr}_M < 90$, when according to the results [52], a sharp increase in ξ is observed

$$\xi/\xi_{\mathrm{tube}} = (1 + 3.1 \cdot 10^6 \, \mathrm{Fr}_M^{-3.4}) \tag{4.92}$$

use of relation (4.77) considerably underestimates the value of the coefficient ξ (Fig. 4.15). The data [43] at $\mathrm{Fr}_M = 56$ are consistent with (4.92). The experimental data [57] for a 7-rod bundle at $\mathrm{Fr}_M = 220$ and $\mathrm{Re} = 10^4$ well agree with (4.86), and at $\mathrm{Fr}_M = 860$ and 1900 these lie much below relation (4.86). Such a behaviour of the relation $\xi/\xi_{\mathrm{tube}} = f(\mathrm{Fr}_M)$ at $\mathrm{Re} = 10^4$ for bundles of spirally finned rods is probably bound up with the specific pattern of flow in such bundles [57]. At the same time the data [4] for bundles of spirally finned rods well agree with (4.86) and (4.92).

Thus, relation (4.90) may be recommended to calculate hydraulic resistance coefficients in helical tube bundles with $\mathrm{Fr}_M \geqslant 80$. At $\mathrm{Fr}_M < 80$ the coefficient $\xi = 3 \, \xi_{\mathrm{tube}}$ may be used for practical calculations.

Analysis of the reported data of various authors has enabled one to give recommendations on calculations of heat transfer and hydraulic resistance coefficients for helical tube bundles.

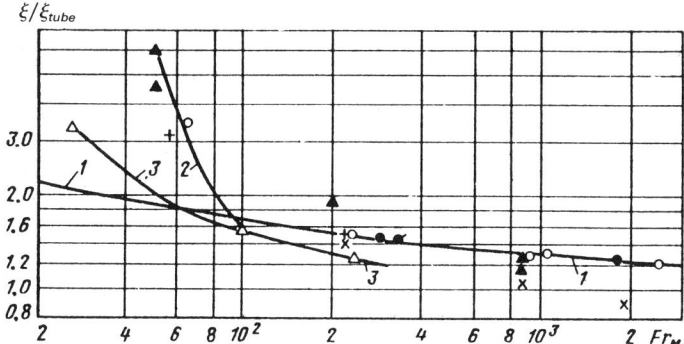

Figure 4.15 Relative hydraulic resistance coefficient as a function of Fr_M at $Re = 10^4$; *1, 2*) relations (4.86) and (4.92); *3*) relation (4.77); ξ based on ξ_{tube}; ○, experimental data [39, 10] at $N \geq 37$; ●, the same at $N = 19$ [39, 51]; ◆, the same at $N = 127$ [39]; △, the same at $N = 19$ [39, 51]; ×, the same at $N = 7$ [57]; +, the same at $N = 127$ [43]; ▲, the same [4].

4.4 LOCAL HEAT TRANSFER AND HYDRAULIC RESISTANCE AT NONUNIFORM HEAT SUPPLY OVER THE BUNDLE CROSS-SECTION

Since the possibility to apply the experimental laws for heat transfer involving uniform heating of a helical tube bundle to the one at nonuniform heat supply to heat carrier in the bundle cross-section is not obvious, a special study has been made of heat transfer with modelling of axisymmetric nonuniformity of heat supply over a bundle radius. So, the simplest case of axisymmetric nonuniformity was analyzed when heat was step-by-step supplied to a group of central helical tubes of a bundle, with electric current being conducted in them, while a peripheral group of helical tubes was not heated. A nonuniform field of heat supply initiated a nonuniform temperature field of heat carrier in a helical tube bundle over the bundle radius which was partially levelled due to cross mixing of the flow. In this case, different conditions are established for heat release from heated helical tubes over the bundle radius and length.

Thus, bundle radius distributions of helical tube wall temperature, specific heat flux, heat carrier temperature and flow velocity outside the wall layer in a given cross-section must be measured to determine local heat transfer coefficients in the considered flow case. Experimental fields of a heat carrier temperature and a flow velocity in the cross-section of a bundle well conform with design temperature and velocity fields determined in the framework of the homogenized flow model. According to this model, the tube and shell wall are conventionally built up with a material layer δ^* in thickness, and free slip flow of a homogenized medium with distributed sources of volumetric energy release and hydraulic resistance is considered for new boundaries of a tube bundle. As this study is based on such a flow model, it is convenient to find the heat transfer coefficient in terms of the parameters at the external wall layer boundary $\bar{T}$ and $\bar{u}$ using the expression [15]:

$$\alpha_m = q_w/[\rho_m \bar{u} c_{pm}(T_w - \bar{T})] \tag{4.93}$$

which is related to the characteristic similarity numbers [10] as:

$$\alpha_m = \alpha_m (Z, Z_m, \text{Pr}_m) \tag{4.94}$$

In relation (4.94) the quantities

$$Z = \frac{\text{Re}_\vartheta}{\beta} = \frac{\bar{u}}{\beta\mu} \int_0^\delta \rho \frac{u}{\bar{u}} (1 - \frac{u}{\bar{u}}) dy \tag{4.95}$$

$$Z_m = \frac{\text{Re}_\Theta}{\alpha_m} = \frac{\bar{u}}{\alpha_m \mu} \int_0^\delta \rho \frac{u}{\bar{u}} \frac{T - \bar{T}}{T_w - \bar{T}} dy \tag{4.96}$$

are the specially constructed Re numbers [15]. The Re_ϑ number in expression (4.95) is determined in terms of the momentum thickness ϑ in the wall layer and the Re_Θ number in (4.96), in terms of the energy thickness Θ.

Since velocity and temperature field distributions on the external part of the wall in a helical tube bundle are described by the same logarithmic distribution laws and the thickness of thermal and dynamic wall layers are equal [10], an assumption may be made $Z = Z_m$. Therefore, in processing the experimental data on heat transfer at nonuniform heat supply over the bundle cross-section, relation (4.94) will be sought in the form:

$$\alpha_m = \alpha_m (Z_m, \text{Pr}_m) \tag{4.97}$$

Studies of heat transfer and the hydraulic resistance coefficient were made on experimental set-ups equipped with bundles of 37 and 127 helical tubes covered with electric-conducting varnish and with air as heat carrier. On these set-ups, electric energy was supplied to the 7th and 37th central helical tubes to promote a stepwise axisymmetric heat release distribution over the bundle radius. These set-ups were detailed in [39]. Study was made in the following range of the parameters: $\text{Fr}_M = 57-1082$, $\text{Re}_f = (0.05-21) \cdot 10^4$, $q_w = (1.2-1.9) \cdot 10^4$ W/m^2, $T_w \leq 780$ K, $T_f = 460-610$ K, $T_w/T_f < 1.45$. Limiting errors in determining the coefficient α_m were $\pm 15\%$ and the coefficient ξ, $\pm 9\%$ [10].

Velocity and temperature field distributions of heat carrier and helical tube wall over the bundle radius were, first of all, measured for each experimental regime typical of the Re_f number and q_w to determine then the coefficient α_m. Typical experimentally measured distributions of $\bar{u}$, $\bar{T}$ and T_w over the radius of a bundle with a number of tubes equal to 127 are shown in Fig. 4.16a,b. Knowing these distributions for a given bundle ra-

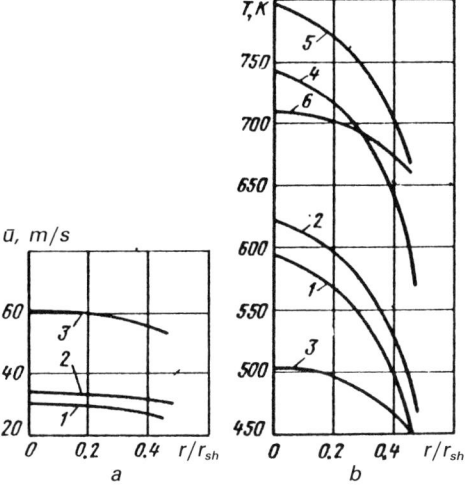

Figure 4.16 Bundle radius distribution of a flow core velocity (*a*), heat carrier and tube wall temperatures (*b*) using the homogeneized flow model: *1–3*) experimental data for $Fr_M = 57$, 236, 1082, respectively; *4–6*) experimental data for a helical tube wall temperature.

dius allows calculation of α_m and Z_m as follows. A dimensionless heat transfer coefficient

$$\bar{\alpha} = q_w/(T_w - \bar{T}) \tag{4.98}$$

is introduced and similarity numbers

$$\bar{Re}_{\delta f} = \bar{u}\bar{\rho}\delta/\mu_f \tag{4.99}$$

$$\bar{Re}_{\delta m} = \bar{u}\bar{\rho}\delta(\bar{T}/T_m)/\mu_m \tag{4.100}$$

$$\bar{Nu}_{\delta m} = \bar{\alpha}\delta/\lambda_m \tag{4.101}$$

are determined. Here

$$\bar{\rho} = p/R\bar{T},\ \mu_f = \mu(\bar{T}),\ T_m = (T_w + \bar{T})/2$$

$$\mu_m = \mu(T_m),\ \lambda_m = \lambda(T_m) \tag{4.02}$$

A mean-integral wall layer thickness in (4.99)–(4.101) is determined by the expression

$$\delta = 0.5(1 + 3.6 Fr_M^{-0.357})^{-4} d_{eq} \tag{4.103}$$

In [10, 39], it is shown that the hydraulic resistance coefficient does not depend on a value of the temperature factor T_w/T_f. A dimensionless friction coefficient may be, therefore, determined by the formula:

$$\beta = 0.266(1 - 4\delta^*/d_{eq})^{1.75}/(8\bar{Re}_{\delta f}^{0.25})$$
or
$$\beta = 0.045\,Z_m^{-0.221} + 4 \cdot 10^{-4} \tag{4.104}$$

where in accord with [15], the boundary layer displacement thickness δ^* may be found by the relation:

$$\delta^* = 1.3 Z_m \delta \mu_m \beta / (\bar{Re}_{\delta f} \mu_f) \tag{4.105}$$

Then expression (4.96) for the quantity Z_m may be reduced to the form

$$Z_m = \bar{Re}_{\delta m}(1 - 2\sqrt{\beta}/0.39)/(0.39\sqrt{\beta}) \tag{4.106}$$

In determining relation (4.104), the following relationship between the maximum velocity $\bar{u}$ (flow core velocity) and mean-mass velocity u_{mean} was taken into account:

$$u_{\text{mean}}/\bar{u} = 1 - 4\delta^*/d_{eq} \tag{4.107}$$

An unknown value of α_m, (4.97), may be expressed in terms of $\tilde{Nu}_{\delta m}$, (4.101), and $\bar{Re}_{\delta m}$, (4.100), as follows

$$\alpha_m = \tilde{Nu}_{\delta m} / (\bar{Re}_{\delta m} Pr_m) \tag{4.108}$$

The experimental data on the heat transfer coefficient α_m as a function of characteristic number Z_m are plotted in Fig. 4.17 and are compared with the relation

$$\alpha_m = (30.4 Z_m^{0.174} Pr_m + 14.65 Z_m^{0.09} - 11.2)^{-1} \tag{4.109}$$

obtained for uniform heating of helical tubes over the bundle radius [10]. Coincidence of the experimental data on the coefficient α_m for $Fr_M = 230–1052$ at nonuniform heat supply with relation (4.109) is good. The experimental values of the coefficient α_m at $Fr_M = 64$ lie approximately by 20% above relation (4.109), which is consistent with the results [39]. It is apparent that at $Fr_M < 90$ there occurs additional flow turbulization due to flow separation from the spiral tube surfaces, which is not observed over the range of $Fr_M > 90$.

Thus, the proposed methods of experimental study of local heat transfer

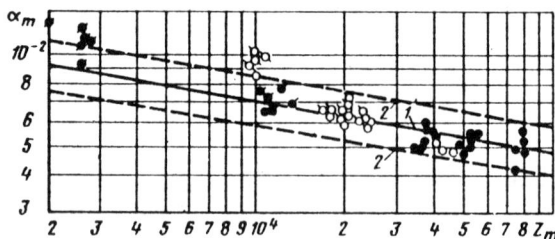

Figure 4.17 Dimensionless heat transfer coefficient in a helical tube bundle vs specially constructed Re number: ●, $Fr_M = 1082$; $L = 0.5$ m; ◆, 236 and 0.5 m; ◆, 57 and 0.5 m; ○, 1050 and 0.75 m; ◒, 232 and 0.75 m; ◓, 57 and 0.75 m; *1*) relation (4.109); *2*) lines of heat transfer coefficient deviations.

in helical tube bundles offer, within a sufficient accuracy, determining the heat transfer coefficients at nonuniform heat supply to heat carrier over a bundle radius. The obtained results for the coefficient α_m point to the fact that the homogenized flow model can be adopted to calculate heat transfer in terms of local flow characteristics using the heat transfer law, (4.109). In this case, a wall layer thickness-mean temperature and a velocity at the external boundary of a wall layer (in the flow core) are taken as characteristic parameters.

The hydraulic resistance coefficient at nonuniform heat supply over a bundle cross-section was studied by the methods cited in [39]. In addition, a mean value of ξ along the control section was calculated by the formula

$$\xi = \frac{\Delta p - \left(\dfrac{G}{F_f}\right)^2 \left[\left(\dfrac{1}{\rho_2}\right) - \left(\dfrac{1}{\rho_1}\right)\right]}{(l_{control}/d_{eq})(G/F_f)^2(1/\rho_{mean})} \qquad (4.110)$$

Experimental data on ξ for a bundle with $Fr_M = 1050$ at the length $l_{control}/d_{eq} = 88$ at nonuniform heat release over a bundle cross-section $q_w = \text{var}(F)$ are plotted in Fig. 4.18 and compared with relation (4.90) obtained at uniform heat release over a bundle cross-section ($q_w = \text{const}(F)$). This figure also shows the experimental data for $q_w = \text{const}(F)$ and $q_w = 0$. It is seen that the experimental data for $q_w = \text{var}(F) = \text{const}(F)$ and $q_w = 0$ well agree with relation (4.90) exhibiting no contribution of different heat carrier temperature distributions over a bundle radius to the coefficient ξ. Similar results are obtained for bundles with $Fr_M = 232$ and 64. Thus, formulas (4.90) and (4.104) obtained at $q_w = \text{const}(F)$ may be employed to calculate hydraulic resistance coefficients at nonuniform heat release in a bundle cross-section.

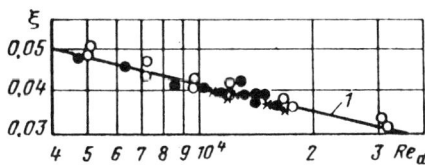

Figure 4.18 Hydraulic resistance coefficient for a helical tube bundle at $Fr_M = 1050$ with nonuniform heat supply vs Re_d number: *1)* relation (4.90); *, ○, ●, experimental data for nonuniform heat supply, nonisothermal heating and adiabatic heating, respectively.

CHAPTER
FIVE

TRANSPORT PROPERTIES OF FLOW AND METHODS OF CALCULATING UNSTEADY HEAT AND MASS TRANSFER

5.1 THEORETICAL CALCULATION METHODS OF UNSTEADY TEMPERATURE FIELDS IN A HELICAL TUBE BUNDLE

The flow model for a homogenized medium which replaces a real bundle of helical tubes and is reliable when applied to calculating steady temperature fields is used to study both unsteady heat and mass transfer under the conditions of a nonuniform heat release field in a bundle cross-section and to determine effective turbulent diffusion coefficients. A homogenized medium consists of heat carrier and solid phase. However, if in the case of a steady process use is made of a one-temperature flow model when calculation of the system of equations (1.8)–(1.11) yields a heat carrier temperature distribution alone, then in the unsteady-state case account is made of thermal inertia of helical tubes by a two-temperature model which also allows for a time variation of solid phase temperature. For this case, the system of the hydrodynamics and energy equations for a gas flow and the heat conduction equation for a solid phase (1.36)–(1.40) is solved.

When perturbations of the parameters specifying a flow pattern are not large and a perturbation time exceeds the sonic wave propagation one along a bundle, gasdynamics equations (1.38)–(1.39) may be written to a quasi-stationary approximation, using, instead of continuity equation (1.39), the relation for heat carrier flow rate in the form [8, 28]

$$G(\tau) = 2\pi m \int_0^{r_{sh}} \rho u r \, dr \qquad (5.1)$$

the function $G(\tau)$ being a known quantity. Motion equation (1.38) to a quasi-stationary approximation is of the form

$$\rho u \frac{\partial u}{\partial x} = -\frac{\partial p}{\partial x} - \xi \frac{\rho u^2}{2 d_{eq}} + \frac{1}{r}\frac{\partial}{\partial r}\left(r \rho \nu_{eff} \frac{\partial u}{\partial r} \right) \qquad (5.2)$$

The coefficients λ_{eff}, ν_{eff}, ξ, α or the coefficients $K = D_t/ud_{eq}$, α and ξ found from experiment were used to close this system of equations.

In the system of equations (1.36)–(1.40), heat conduction equation for a solid phase (1.36) was written for the case if the thermal conductivity of a solid phase λ_s might be taken as the isotropic one not depending on a direction. Indeed, this coefficient depends on coordinates. Then equation (1.36) assumes the form

$$\rho_s c_s \frac{\partial T_s}{\partial \tau} = q_v - \frac{4\alpha m}{(1-m)d_{eq}}(T_s - T)$$
$$+ \frac{1}{r}\frac{\partial}{\partial r}\left(r\lambda_{sr} \frac{\partial T_s}{\partial r} \right) + \frac{\partial}{\partial x}\left(\lambda_{sx} \frac{\partial T_s}{\partial x} \right) \qquad (5.3)$$

where λ_{sr} is the thermal conductivity of a solid phase in the radial direction and λ_{sx}, the same in the longitudinal direction.

Owing to the fact that a solid phase is uniformly distributed over the volume of the shell, into which a helical tube bundle is placed, the solid phase temperature T_s is mainly a helical tube surface temperature at a fixed spatial point at a given time moment τ. Knowing a temperature distribution T_s over the outer surface of real tubes, it is possible to determine unsteady temperature fields in a tube wall under the 3rd-kind boundary conditions on the inner side of tubes. However, in solving the system of equations (5.1)–(5.3), (1.37), (1.40) a condition must be required for the same thermal inertia properties both of a real bundle and of a solid phase in a homogenized medium. Only in this case, it is possible to attain coincidence of experimentally measured and predicted temperature fields of heat carrier and a solid phase.

In experiments made on the set-ups described in Chapter 5, air was stagnant inside the tubes and a tube wall thickness was 0.2–0.5 mm. Under these conditions convective heat transfer inside the tubes was absent. Therefore, in determining the quantities ρ_s, c_s, λ_{sr}, λ_{sx}, it has appeared possible to adopt the following design schemes.

So, in calculating ρ_s, account was taken of a volume occupied by heat carrier flowing inside the tubes. Then

$$\rho_s = \rho_{max}(1 - \epsilon) + \rho_g \epsilon \tag{5.4}$$

where ϵ is the flow-to-total cross-sectional tube area ratio.

Thermal conductivities of a solid phase in the radial and longitudinal directions λ_{sr} and λ_{sx}, respectively, in a general case must allow for thermal resistances at the points of contacting the tubes. As there are no experimental data on these coefficients, in calculating temperature fields of heat carrier in helical tube bundles, the following relations are used

$$\lambda_{sr} = \left[\frac{1-\epsilon}{\lambda_{mat}} + \frac{\epsilon}{\lambda_g}\right]^{-1} \tag{5.5}$$

$$\lambda_{sx} = \lambda_{mat}(1 - \epsilon) + \lambda_g \epsilon \tag{5.6}$$

where λ_{mat} and λ_g is the thermal conductivity of the tube material and heat carrier, respectively. These relations allow for the impact of the designing features of helical tubes on heat transfer by conduction in a solid phase and are obtained using the concept of equivalent thermal conductivity of a multilayer wall composed of different materials (wall and heat carrier). As numerical experiments with varying λ_{sr} and λ_{sx} have shown, such an approach is quite applicable in the framework of the homogenized flow model.

Heat capacity of a solid phase has been found by the formula

$$c_s = c_{max}(1 - \epsilon) + c_g \epsilon \tag{5.7}$$

The system of equations (5.1)–(5.3), (1.37), (1.40) is solved by the numerical method and is supplemented with the following boundary conditions:

$$T_s(r, 0, \tau) = T_{s.in}(r, \tau) \tag{5.8}$$

$$T(r, 0, \tau) = T_{in}(r, \tau) \tag{5.9}$$

$$u(r, 0, \tau) = u_{in}(r, \tau) \tag{5.10}$$

$$p(r, 0, \tau) = p_{in}(\tau) \tag{5.11}$$

at the bundle outlet (no heat transfer)

$$\frac{\partial T_s(r, x, \tau)}{\partial x}\bigg|_{x=1} = 0, \quad \frac{\partial T(r, x, \tau)}{\partial x}\bigg|_{x=1} = 0 \tag{5.12}$$

on the bundle axis (axial symmetry)

$$\frac{\partial T_s(r, x, \tau)}{\partial r}\bigg|_{r=0} = 0, \quad \frac{\partial T(r, x, \tau)}{\partial r}\bigg|_{r=0} = 0 \tag{5.13}$$

$$\left.\frac{\partial u}{\partial r}\right|_{r=0} = 0 \tag{5.14}$$

at the external boundary of a bundle

$$-\lambda_{sr}\left.\frac{\partial T_s(r, x, \tau)}{\partial r}\right|_{r=r_{sh}} = 0, \quad -\lambda_{eff}\left.\frac{\partial T(r, x, \tau)}{\partial r}\right|_{r=r_{sh}} = 0$$

$$\left.\frac{\partial u}{\partial r}\right|_{r=r_{sh}} = 0 \tag{5.15}$$

Initial conditions are found by solving the steady-state problem at the time moment $\tau = 0$. In solving system (5.1)–(5.3), (1.37), (1.40) with boundary conditions (5.8)–(5.15), the quantities at the derivatives were preliminarily averaged depending on the differentiation coordinates and were removed from the differentiation sign and then refined in iteration cycles.

Heat conduction and energy equations were solved by the method of variable directions [34]. Numerical analogs of the equations in this case were broken down according to an implicit scheme and were solved by the elimination method. The latter based on using Simuni's substitution was adopted to solve the motion and continuity equations. Thus, solving the problem was divided into two successive stages: solving heat conduction equations (1.37) and (5.3) and simultaneously solving motion and continuity equations (5.1) and (5.2) which were then matched in terms of state equation (1.40) and the iteration cycles.

A network r_i, x_i, τ^ν with steps Δr, Δx and $\Delta \tau$ was chosen to numerically solve the systems of equations (5.1), (5.3), (1.37) and (1.40). A time step was chosen according to the relation:

$$\Delta \tau / \Delta x \leq 1/(u + a) \tag{5.16}$$

obtained by solving a simplified linear system of differential gasdynamics equations. In formula (5.16), a is the sonic velocity. Since disturbance waves in a gas propagate with a sonic velocity, it may be assumed that unsteady gasdynamic processes in a helical tube bundle proceed for 0.1 s. This offers calculation of relatively "slow" processes as quasi-stationary with a time step $\Delta \tau$ which is greater than the propagation time of weak perturbations. In solving the thermal problem, there are no stringent restrictions on a choice of a time step since use is made of the method of variable directions together with an implicit scheme being stable over a wide range of spatial-time steps. A solution algorithm has been implemented in the form of the FORTRAN computer program for BESM-6 [32].

The aforesaid calculation method assumes that a velocity vector is parallel to the tube bundle axis. However, there may exist cases of flow in a bundle, when, in calculating unsteady heat and mass transfer, a radial velocity component must be allowed for. For this, it is possible to employ the calculation method based on a two-temperature flow model for a two-phase homogenized medium involving a fixed solid phase, and the tube bundle flow may, with regard to volumetric energy release and friction sources, be described by the following initial system of equations [8]:

$$\rho_s c_s \frac{\partial T_s}{\partial \tau} = q_v - \frac{4\alpha m}{(1-m)d_{eq}}(T_s - T) + \frac{1}{r}\frac{\partial}{\partial r}\left(r\lambda_{sr}\frac{\partial T_s}{\partial r}\right)$$

$$+ \frac{\partial}{\partial x}\left(\lambda_{sx}\frac{\partial T_s}{\partial x}\right) \tag{5.17}$$

$$\rho c_p \frac{\partial T}{\partial \tau} + \rho u c_p \frac{\partial T}{\partial x} + \rho v c_p \frac{\partial T}{\partial r} = \frac{4\alpha}{d_{eq}}(T_s - T)$$

$$+ \frac{1}{r}\frac{\partial}{\partial r}\left(r\lambda_{eff}\frac{\partial T}{\partial r}\right) + \frac{\partial}{\partial x}\left(\lambda_{eff}\frac{\partial T}{\partial x}\right) \tag{5.18}$$

$$\rho u \frac{\partial u}{\partial x} + \rho v \frac{\partial u}{\partial r} = -\frac{\partial p}{\partial x} - \xi \frac{\rho u^2}{2d_{eq}} + \frac{1}{r}\frac{\partial}{\partial r}\left(\rho r v_{eff}\frac{\partial u}{\partial r}\right) \tag{5.19}$$

$$\frac{\partial(\rho u m)}{\partial x} + \frac{1}{r}\frac{\partial}{\partial r}(r\rho vm) = 0 \tag{5.20}$$

$$p = \rho R T \tag{5.21}$$

In a general case, this system of equations must be supplemented with the motion equation for a radial direction. However, considering the mathematical difficulties associated with solving the system of equations incorporating the motion equation both for the axial and radial directions, in the present book, an assumption has been made that $\partial p/\partial r = 0$, and the methods of numerical calculation of temperature fields have been developed for the system of equations (5.17)–(5.21). The condition $\partial p/\partial r = 0$ in a helical tube bundle is supported by experiment. Equations (5.17)–(5.21) have been supplemented with the following boundary conditions:

(a) at the bundle inlet ($x = 0$) temperature, velocity and pressure profiles are known;
(b) on the bundle axis ($r = 0$):

$$\frac{\partial T_s}{\partial r} = 0, \frac{\partial T}{\partial r} = 0, \frac{\partial u}{\partial r} = 0, v = 0 \tag{5.22}$$

(c) at the boundary ($r = r_{sh}$)

$$\frac{\partial T_s}{\partial r} = 0, \frac{\partial T}{\partial r} = 0, \frac{\partial u}{\partial r} = 0, v = 0 \tag{5.23}$$

(d) at the bundle outlet ($x = l$)

$$\frac{\partial T_s}{\partial x} = 0, \quad \frac{\partial T}{\partial x} = 0 \tag{5.24}$$

Initial distributions $T_s(r, x)$, $T(r, x)$ and $u(r, x)$ at $\tau = 0$ have been found by solving the steady-state problem.

The initial system of equations (5.17)–(5.21) is nonlinear and incorporates the coefficients that depend on a solution and are under the differentiation signs. Such systems may be solved only numerically [6, 35].

The system was preliminarily quasi-linearized before it might be solved. This procedure enabled one to remove the coefficients λ_{sx}, λ_{sr}, λ_{eff} and ν_{eff} from the derivative sign and to refine them in iteration cycles.

To numerically solve the system of equations, a network r_i, x_j, τ^ν with the appropriate network steps Δr, Δx, $\Delta \tau$ was chosen, and it divided the domain of unknown functions into M concentric zones over a radius and into N layers over a height.

The method of variable directions [37] as most applicable to parabolic-type equations was chosen to solve heat transfer equations. It should be noted that as implicit finite difference schemes are used along with the method of variable directions, the latter offers avoiding the stringent restrictions on the space-time step ratio. A pattern of a chosen network is shown in Fig. 5.1.

In accord with the chosen pattern, the heat conduction equations are splitted into two one-dimensional ones in the r-and x-directions. As an example, present the finite-difference analog of equation (5.17):

$$2\bar{\rho}_s \bar{c}_s \frac{T_{si,j}^{\nu+1/2} - T_{si,j}^\nu}{\Delta \tau} = q_{vi,j}^{\nu+1} - \frac{4\alpha m}{(1-m)d_{eq}}(T_{si,j}^{\nu+1/2}$$

$$- T_{i,j}^{\nu+1/2}) + \bar{\lambda}_{sr} \frac{T_{si+1,j}^{\nu+1/2} - T_{si-1,j}^{\nu+1/2}}{r_i \cdot 2\Delta r}$$

$$+ \bar{\lambda}_{sr} \frac{T_{si+1,j}^{\nu+1/2} - 2T_{si,j}^{\nu+1/2} + T_{si-1,j}^{\nu+1/2}}{\Delta r^2}$$

$$+ \bar{\lambda}_{sx} \frac{T_{si,j+1}^\nu - 2T_{si,j}^\nu + T_{si,j-1}^\nu}{\Delta x^2} \tag{5.25}$$

$$2\bar{\rho}_s \bar{c}_s \frac{T_{si,j}^{\nu+1} - T_{si,j}^{\nu+1/2}}{\Delta \tau} = q_{vi,j}^{\nu+1} - \frac{4\alpha m}{(1-m)d_{eq}}(T_{si,j}^{\nu+1} - T_{i,j}^{\nu+1})$$

$$+ \bar{\lambda}_{sr} \frac{T_{si+1,j}^{\nu+1/2} - T_{si-1,j}^{\nu+1/2}}{r_i \cdot 2\Delta r} + \bar{\lambda}_{sr} \frac{T_{si+1,j}^{\nu+1/2} - 2T_{si,j}^{\nu+1/2} + T_{si-1,j}^{\nu+1/2}}{\Delta r^2}$$

$$+ \bar{\lambda}_{sx} \frac{T_{si,j+1}^{\nu+1} - 2T_{si,j}^{\nu+1} + T_{si,j-1}^{\nu+1}}{\Delta x^2} \tag{5.26}$$

CALCULATING UNSTEADY HEAT AND MASS TRANSFER **127**

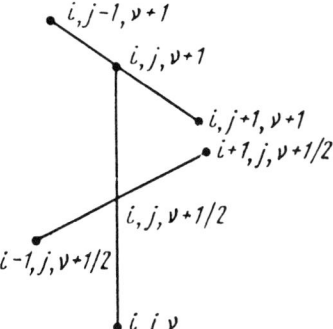

Figure 5.1 Pattern of a design network.

Simuni's substitution was utilized to solve motion equation (5.19). Substituting, into the finite-difference analog, the motion equations $u_{ij} = \bar{u}_{i,j} + \tilde{\bar{u}}_{ij}(\partial p/\partial x)$ yields two following equations

$$\overline{\rho u}\frac{\bar{u}_{i,j} - \bar{u}_{i,j-1} - \tilde{\bar{u}}_{i,j-1}\left(\frac{\partial p}{\partial x}\right)_{j-1}}{\Delta x} + \rho\bar{v}\frac{\bar{u}_{i+1,j} - \bar{u}_{i-1,j}}{2\Delta r}$$

$$= \xi\overline{\rho u}\frac{\bar{u}_{i,j}}{2d_{eq}} + \rho\bar{v}_{eff}\frac{\bar{u}_{i+1,j} - \bar{u}_{i-1,j}}{2r_i\Delta r}$$

$$+ \overline{\rho v}_{eff}\frac{\bar{u}_{i-1,j} - 2\bar{u}_{i,j} + \bar{u}_{i+1,j}}{\Delta r^2} \qquad (5.27)$$

$$\overline{\rho u}\frac{\tilde{\bar{u}}_{i,j}}{\Delta x} + \overline{\rho v}\frac{\tilde{\bar{u}}_{i+1,j} - \tilde{\bar{u}}_{i-1,j}}{2\Delta r} = -1 - \xi\overline{\rho u}\frac{\tilde{\bar{u}}_{i,j}}{2d_{eq}}$$

$$+ \overline{\rho v}_{eff}\frac{\tilde{\bar{u}}_{i+1,j} - \tilde{\bar{u}}_{i-1,j}}{2r_i\Delta r} + \overline{\rho v}_{eff}\frac{\tilde{\bar{u}}_{i+1,j} - 2\tilde{\bar{u}}_{i,j} + \tilde{\bar{u}}_{i-1,j}}{\Delta r^2} \qquad (5.28)$$

where $\bar{u}_{i,j}$ and $\tilde{\bar{u}}_{i,j}$ are some components of the longitudinal velocity vector component.

A pressure gradient is found from the relation

$$\frac{\partial p}{\partial x} = \frac{G - 2\pi m\int_0^{r_{sh}}\rho\bar{u}rdr}{2\pi m\int_0^{r_{sh}}\rho\tilde{\bar{u}}rdr} \qquad (5.29)$$

Solving continuity equation (5.20) does not involve great difficulties as a value of the radial velocity vector component v on a new design layer is evaluated from the recurrence formulas. Calculations have evidenced that at a constant porosity and at a constant hydraulic diameter the value of v is close to zero.

In this case, solving the stated problem was also divided into two successive stages: "thermal" part of the problem (equations (5.17) and (5.18)) and "gasdynamic" (equations (5.19) and (5.20)). Solutions of both parts are matched through state equation (5.21).

It should be noted that the present calculation methods provide for possible variations of the porosity and hydraulic diameter both with respect to a radius and with respect to a height, i.e. $m = f(r, x)$ and $d_{eq} = \psi(r, x)$.

Following the cited methods, an algorithm is constructed and a FORTRAN computer program is composed, as applied to BESM-6 [8].

An experimentally measured dimensionless unsteady effective turbulent diffusion coefficient

$$K_{uns} = D_t/ud_{eq}$$

has been used to close the system of equations (5.17)–(5.21) as it is related to the effective turbulent thermal conductivity $\lambda_{eff} = D_t \rho c_p$ and kinematic viscosity coefficient $\nu_{eff} = D_t$, assuming that the turbulent Prandtl and Lewis numbers are equal to unity.

The aforesaid methods to calculate the unsteady axisymmetric problem for heat and mass transfer in helical tube bundles offer calculating heat carrier temperature fields at prescribed nonuniformity of a heat release (heat supply) field. Comparing these theoretical temperature fields with those measured experimentally on the helical tube bundle models allows effective turbulent diffusion coefficients K_{uns} to be determined at different time instants and the effect of different performance parameters on these coefficients to be studied.

In a number of cases, asymmetric nonuniformity of heat release (heat supply) may be realized in the cross-section of a helical tube bundle, e.g. at lateral supply of heat carrier to the bundle inlet and its removal from the bundle outlet. Then the solution method [36] may be adopted.

A system of equations for unsteady flow in a helical tube bundle, in the homogenized flow statement for asymmetric nonuniformity of an energy release field, incorporates equations of motion, continuity, energy, state and heat conduction which describe a temperature distribution in helical tubes (bundle "skeleton"). In this case, the gasdynamics equations may be written, to a quasi-stationary approximation, by replacing the continuity equation by the flow rate one at increasing or decreasing heat power when a heat carrier flow rate does not vary in time:

$$\rho_s c_s \frac{\partial T_s}{\partial \tau} = q_v(r, \varphi, x, \tau) - \frac{4\alpha m}{(1-m)d_{eq}}(T_s - T)$$

$$+ \frac{1}{r}\frac{\partial}{\partial r}\left(r\lambda_{sr}\frac{\partial T_s}{\partial r}\right) + \frac{\partial}{\partial x}\left(\lambda_{sx}\frac{\partial T_s}{\partial x}\right) + \frac{1}{r^2}\frac{\partial}{\partial \varphi}\left(\lambda_{s\varphi}\frac{\partial T_s}{\partial \varphi}\right) \quad (5.30)$$

$$\rho c_p \frac{\partial T}{\partial \tau} + \rho u c_p \frac{\partial T}{\partial x} = \frac{4\alpha}{d_{eq}}(T_s - T) + \frac{1}{r}\frac{\partial}{\partial r}\left(r\lambda_{eff}\frac{\partial T}{\partial r}\right)$$

$$+ \frac{\partial}{\partial x}\left(\lambda_{eff}\frac{\partial T}{\partial x}\right) + \frac{1}{r^2}\frac{\partial}{\partial \varphi}\left(\lambda_{eff}\frac{\partial T}{\partial \varphi}\right) \quad (5.31)$$

$$\rho u \frac{\partial u}{\partial x} = -\frac{\partial p}{\partial x} - \xi\frac{\rho u^2}{2d_{eq}} + \frac{1}{r}\frac{\partial}{\partial r}\left(r\rho\nu_{eff}\frac{\partial u}{\partial r}\right)$$

$$+ \frac{1}{r^2}\frac{\partial}{\partial \varphi}\left(\rho\nu_{eff}\frac{\partial u}{\partial \varphi}\right) \quad (5.32)$$

$$G = m\int_0^{2\pi}\int_0^{r_{sh}} \rho u r\, dr\, d\varphi \quad (5.33)$$

$$p = \rho R T \quad (5.34)$$

Boundary conditions of the problem are:
at the initial time moment ($\tau = 0$)

$$T_s(r, \varphi, x, 0) = T_{s0}(r, \varphi, x)$$

$$T(r, \varphi, x, 0) = T_0(r, \varphi, x) \quad (5.35)$$

$$u(r, \varphi, x, 0) = u_0(r, \varphi, x)$$

$$p(x, 0) = p_0(x)$$

conditions at the bundle inlet ($x = 0$)

$$u(r, \varphi, 0) = u_{in}(r, \varphi)$$

$$T(r, \varphi, 0, \tau) = T_{in}(r, \varphi, \tau) \quad (5.36)$$

$$T_s(r, \varphi, 0, \tau) = T_{s.in}(r, \varphi, \tau)$$

azimuthal periodicity conditions

$$u(r, \varphi, x) = u(r, \varphi + 2\pi, x)$$

$$T(r, \varphi, x, \tau) = T(r, \varphi + 2\pi, x, \tau) \quad (5.37)$$

$$T_s(r, \varphi, x, \tau) = T_s(r, \varphi + 2\pi, x, \tau)$$

conditions at the bundle boundaries

at $x = l$ $\quad \dfrac{\partial T_s}{\partial x}\bigg|_{x=1} = 0; \quad \dfrac{\partial T}{\partial x}\bigg|_{x=1} = 0 \quad (5.38)$

at $r = r_{sh}$ $\quad \dfrac{\partial T_s}{\partial r}\bigg|_{r=r_{sh}} = 0; \quad \dfrac{\partial T}{\partial r}\bigg|_{r=r_{sh}} = 0; \quad \dfrac{\partial u}{\partial r}\bigg|_{r=r_{sh}} = 0 \quad (5.39)$

Since in [16] it was shown that in the steady-state case the coefficients D_t used to close systems of the equations describing both asymmetric and axisymmetric problems were identical, it may be assumed that the same situation will be also observed in the unsteady-state case. Therefore, experimental determination of the coefficient D_t may be confined to the case of axisymmetric nonuniformity of heat release.

This problem is solved by the method of variable directions. In this case, the initial conditions are splitted into three directions, and then a solution to this system [36, 38] may be found by the two-layer iteration methods.

5.2 UNSTEADY MIXING INVOLVING A SHARP CHANGE IN HEAT POWER

Experimental study of unsteady heat and mass transfer in a helical tube bundle at $Fr_M = 220$ was made on the experimental set-up detailed in Chapter 2. Moreover, measurements were made of heat carrier temperature fields in the intertube space in the flow core, which is consistent with the accepted flow model for a homogenized medium and enables one to compare experimental and predicted temperature fields as well as to determine unsteady effective turbulent diffusion coefficient. Heat carrier temperature fields were measured in the outlet cross-section of a helical tube bundle by a comb composed of 10 chromel-alumel 0.1 mm dia wire thermocouples mounted at the cell centres with the coordinates $r/r_{sh} = 0.073, 0.128, 0.193, 0.265, 0.334, 0.408, 0.479, 0.624, 0.770, 0.916$. Time variations of heat power and temperature at sharply increasing and decreasing heat load are plotted in Fig. 5.2 at $Re = 8.9 \cdot 10^3$ and $1.75 \cdot 10^4$, respectively.

Figure 5.3a,b,c plots experimentally measured air temperature fields at different time instants ($\tau = 4, 6, 8, 10, 12, 20, 30, 32, 36$ s) with sharply increasing heat release at $Re = 8.9 \cdot 10^3, 1.36 \cdot 10^4$ and $1.75 \cdot 10^4$. Time increase of heat release has been made at a constant heat carrier flow rate. Figure 5.3a,b,c also plots predicted heat carrier temperature fields at the same time instants but at different values of the coefficients $K = D_t/ud_{eq}$. In comparing experimental and predicted temperature fields of heat carrier, good agreement between these distributions is observed but at different values of the coefficient K for different time instants. It is seen too that this coefficient decreases with increasing time within the interval $\tau = 0$–10 s.

The modified least square method was adopted to determine numerical values of the coefficient K at each time instant. In this case, quite a certain value of the coefficient K is assigned on the curves $T = T(r, K, \tau)$ (Fig. 5.3a,b,c) according to the plotted grid of the theoretical curves, and the squares of deviations of each point δ_i^2 from these curves at prescribed values of K are calculated. Then the relation

$$\sqrt{\sum_{i=1}^{n} \delta_i^2} = f(K) \tag{5.40}$$

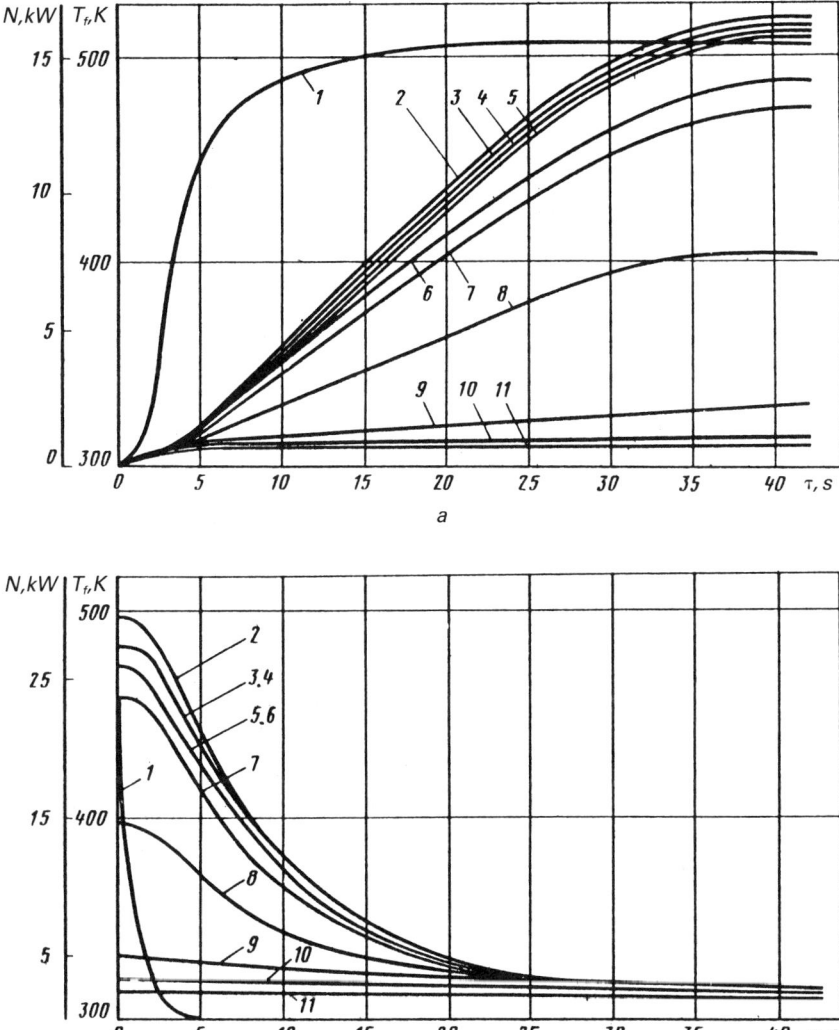

Figure 5.2 Typical time variations of power and temperature of heat carrier at sharply increasing heat power at $Re_b = 8.9 \cdot 10^3$ (*a*) and at sharply decreasing heat power at $Re_b = 1.75 \cdot 10^4$ (*b*): *1*) heat power; *2–11*) temperature vatiation at $r/r_{sh} = 0.073, 0.128, 0.193, 0.265, 0.334, 0.408, 0.479, 0.624, 0.770, 0.916$, respectively.

is plotted where n is the number of experimental points, and a minimum of function (5.40) is evaluated and corresponds to a maximum confidence value of K, at which the best coincidence between experimental and predicted temperature fields of heat carrier is provided.

The coefficient K_{uns} determined in this manner is a very complex function of time (Fig. 5.4) which well governs all the experimental data for

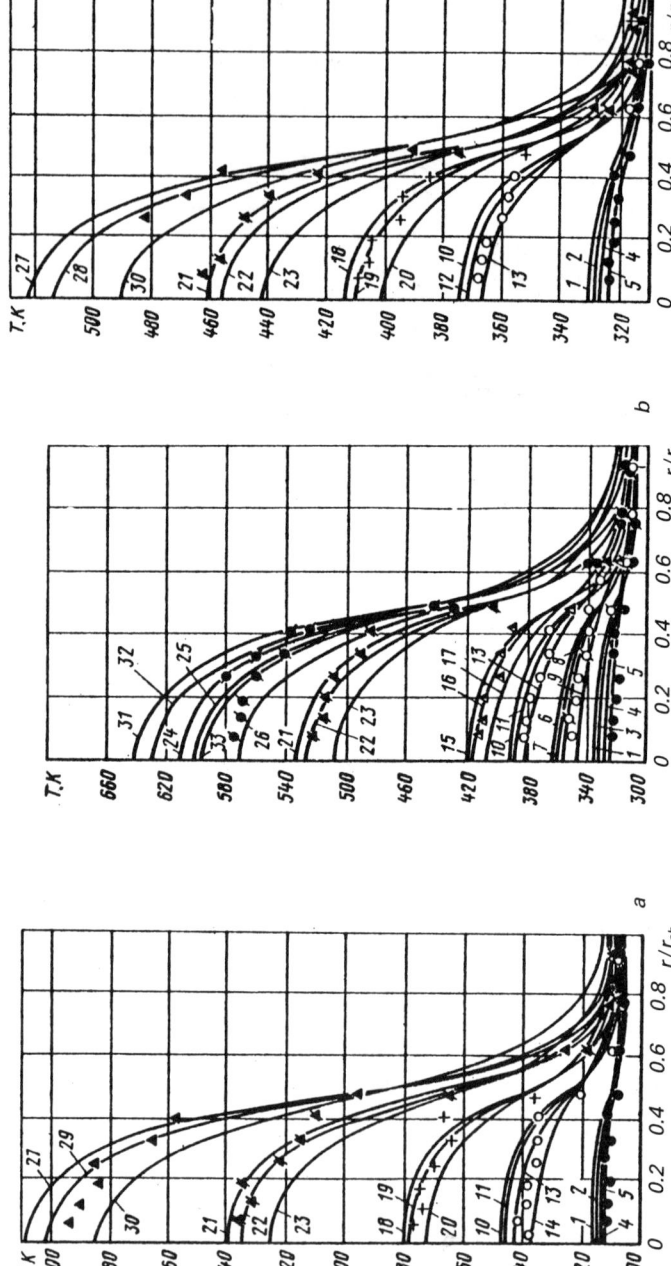

Figure 5.3 Heat carrier temperature fields in the outlet cross-section of a helical tube bundle at $Re_b = 8.9 \cdot 10^3$ (*a*), $Re_b = 1.36 \cdot 10^4$ (*b*), $Re_b = 1.75 \cdot 10^4$ (*c*) at different time instants: *1–5*) calculation at $\tau = 4$ s for coefficients $K = 0.045, 0.06, 0.08, 0.1, 0.2$, respectively; *6–9*) the same $\tau = 6$ s for $K = 0.045, 0.06, 0.1, 0.2$; *10–14*) the same at $\tau = 8$ s for $K = 0.045, 0.06, 0.08, 0.1, 0.2$; *15–17*) the same at $\tau = 10$ s for $K = 0.045, 0.06, 0.1$; *18–20*) the same at $\tau = 12$ s for $K = 0.045, 0.06, 0.1$; *21–23*) the same at $\tau = 20$ s for $K = 0.045, 0.06, 0.1$; *24–26*) the same at $\tau = 30$ s for $K = 0.045, 0.06, 0.1$; *27–30*) the same at $\tau = 32$ s for $K = 0.045, 0.06, 0.08, 0.1$; *31–33*) the same at $\tau = 36$ s for $K = 0.045, 0.06, 0.1$; ●, ∅, ○, ▲, +, ▴, ♦, ▲, ●, experimental data at $\tau = 4, 6, 8, 10, 12, 20, 30, 32, 36$ s, respectively.

different Re numbers. The coefficient K_{uns} more quickly attains a quasi-stationary value of K_{qs} as against temperatures of a wall and heat carrier (Fig. 5.4) A variation of K_{uns} at initial time instants qualitatively coincides with that of the unsteady heat transfer coefficient in round tubes [24, 26] for the same type of unsteadiness which also more quickly achieves a quasi-stationary value as against temperatures of a wall and heat carrier. An assumption may be made that unsteady mixing in helical tube bundles as well as unsteady heat transfer coefficient in channels is primarily affected by the same transfer mechanisms.

However, in a tube bundle the effect of the Re number over the range $(8.9-17.5) \cdot 10^4$ and of the temperature factor $(T_w/T_f)_{max} = 1-1.37$ on the coefficient K_{uns} is not observed (Fig. 5.4).

The coefficient K_{uns} characterizes heat carrier temeprature field variations in the flow core on the diameter scale of a helical tube bundle when the unsteady heat transfer problem is solved in the homogenized flow statement for a nonuniform heat release field over a bundle radius. That the effective unsteady turbulent diffusion coefficient K_{uns} be generalized, the Fourier number (thermal homogeneity) may be used as it specifies a relationship between a rate of varying a temperature field of heat carrier, its physical properties and its dimensions of the considered flow region

$$\text{Fo}_b = \frac{a\tau}{d_{sh}^2} = \frac{\lambda_b \tau}{c_p \rho_b d_{sh}^2} \tag{5.41}$$

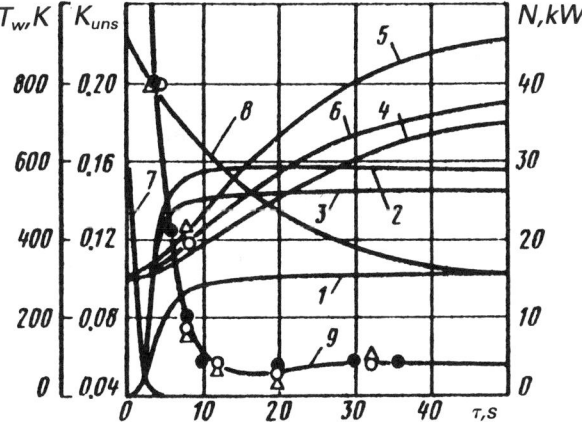

Figure 5.4 Time variations of heat release, tube wall temperature and dimensionless effective turbulent diffusion coefficient: ○, ●, △, experimental data on the coefficient K_{uns} at sharply increasing heat release for $\text{Re}_b = 8.9 \cdot 10^3$, $1.36 \cdot 10^4$, $175 \cdot 10^4$, respectively; *1–3*) time variation of heat release being sharply increased at $\text{Re}_b = 8.9 \cdot 10^3$, $1.36 \cdot 10^4$, $1.75 \cdot 10^4$, respectively; *4–6*) time variation of a wall temperature at heat release being sharply increased at $\text{Re}_b = 8.9 \cdot 10^3$, $1.36 \cdot 10^4$, $1.75 \cdot 10^4$, respectively; *7*) time variation of heat release being sharply decreased at $\text{Re}_b = 1.36 \cdot 10^4$; *8*) time variation of a wall temperature at heat release being sharply decreased at $\text{Re}_b = 1.36 \cdot 10^4$; *9*) coefficient K_{uns} as a function of time.

A functional relation

$$\kappa = K_{uns}/K_{qs} = \kappa\,(Fo_b) \tag{5.42}$$

is found from experiment. It has proved that all the obtained experimental data for K_{uns} at sharply increasing heat release and constant heat carrier flow rate shown in Fig. 5.4 in the form of relation (5.42) are well described by the interpolation relation [27, 28]

$$\kappa = 0.81 \cdot 10^{-4} Fo_b^{-2} - 0.978 \cdot 10^{-2} Fo_b^{-1} + 1.21 \tag{5.43}$$

or

$$K_{uns} = K_{qs}(0.81 \cdot 10^{-4} \cdot Fo_b^{-2} - 0.978 \cdot 10^{-2} Fo_b^{-1} + 1.21) \tag{5.44}$$

where K_{qs} are the quasi-stationary values of the coefficient K determined by the relations obtained in Chapter 4.

Thus, a time variation of the coefficent K_{uns} and $\kappa = K_{uns}/K_{qs}$ (Fig. 5.4) may be, first of all, attributed to varying turbulent flow structure at unsteady heating of a helical tube bundle, thereby reconstructing heat carrier temperature fields. This transfer mechanism is also responsible for the specific features of unsteady heat transfer in channels [24, 26]. Considering that there exists a relationship

$$\alpha = \frac{\lambda(\partial T/\partial r)_{r=r_w}}{T_w - (\int_E \rho u T dF / \int_E \rho u dF)} \tag{5.45}$$

between the heat transfer coefficient α and flow temperature field, the obtained results may be considered as those supporting the hypothesis [24] about the effect of unsteady boundary conditions on the flow structure. Other transfer mechanisms acting upon the coefficient K_{uns} [39, 16, 33] to a lesser degree depend on unsteady boundary conditions. So, the ordered transfer of heat carrier via spiral channels of helical tubes is specified by the relative tube twisting pitch S/d and is not practically affected by the unsteadiness parameters. However, together with convective transfer on the cell scale due to vortex transfer in the cross-section between the wall layer and flow core, it enhances the levelling of temperature field nonuniformity. At the same time, flow swirling in the wall layer following the solid-state law [3] may, to some extent, decrease a turbulence growth at the first time moments. Apparently, this is responsible for the coefficients K_{uns} and K at their maxima (Fig. 5.4) before these attain their quasi-stationary values, which was not observed in studies of unsteady heat transfer in channels. Upon the whole, wall heating augments turbulence generation ($\tau \leq 10$ s), thus reconstructing a temperature field not only in the wall layer, as it was assumed in [24] but also in the flow core (Fig. 5.3). This is produced by the action of all the transfer mechanisms typical of helical tube bundles.

Use of the homogenized flow model, to a certain degree, anticipates a choice of the numbers to be determined in generalizing the data for K_{uns}. Indeed, at nonuniform heat release over a tube bundle radius, a heat carrier temperature field promoted by this bundle is partly levelled, and the values of T_w, T_f and the ratios T_w/T_f are measured over the bundle radius. This results in the ambiguity associated with choosing the characteristic quantities T_w and $\partial T_w/\partial \tau$ in the expression for the unsteadiness criterion [26, 24]:

$$K_{Tg}^* = \frac{\partial T_w}{\partial \tau} \frac{1}{T_w} \sqrt{\frac{\lambda_b}{c_p g \rho_b u}} \qquad (5.46)$$

The coefficient K_{uns} found in the above fashion is a mean quantity for the bundle cross-section where measurements have been made of the distributions of T (Fig. 5.3) at each time instant while the quantities α, u, ρ, etc. vary over the bundle radius. Moreover, in analyzing temperature fields in the homogenized flow statement the scale d_{sh} but not d_{eq} is chosen as a flow scale which is used for one-dimensional description of unsteady heat transfer in a channel.

The results on K_{uns} at sharply decreasing heat release are plotted in Fig. 5.5a,b. As seen from Fig. 5.5, this type of unsteadiness is characterized by decreasing a rate of interchannel mixing at the first time moments, as compared to quasi-stationary operating conditions, which points to a contribution of unsteady boundary conditions to the flow structure, thereby reconstructing heat carrier temperature fields in time and supporting the hypothesis [24]. The variation of the coefficient K_{uns} at this type of unsteadiness coincides with that of the coefficient α in tubes [24, 26]. A criterial relation for calculation of a coefficient in this case is of the form [27]:

$$\kappa = 0.454 \cdot 10^{-5} \text{Fo}_b^{-2} - 3.86 \cdot 10^{-3} \text{Fo}_b^{-1} + 1.24 \qquad (5.47)$$

Experimental data on the coefficient K_{uns} at a sharp increase and at a sharp decrease in heat power are listed in Table 5.1.

By analogy with the methods of generalizing the experimental data on unsteady convective heat transfer in turbulent channel flow [24], it may be assumed that in a general case the experimental data on unsteady mixing coefficient may be represented in the form of the following criterial relation:

$$\kappa = \frac{K_{uns}}{K_{qs}} = \kappa(K_{Tg}^*, K_s, K_G, \text{Re}_f, T_w/T_f) \qquad (5.48)$$

where the dimensionless parameter K_{Tg}^* is determined by (5.46) and allows for a turbulent flow structure change in the wall layer due to a wall temperature variation and, therfore, as shown in [24], it characterizes a variation of thermal resistance between the wall and flow under the considered conditions. Parameter (5.46) in a bundle must take into account the effect of unsteadiness on the thermal resistance between the wall and the flow in the cell of a helical tube bundle as well as on the thermal resistance between

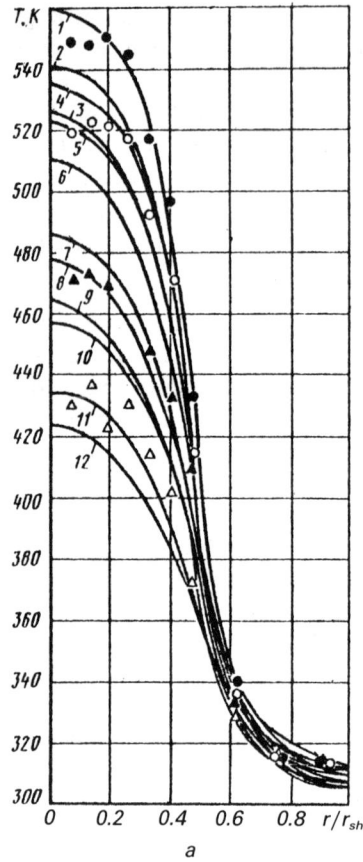
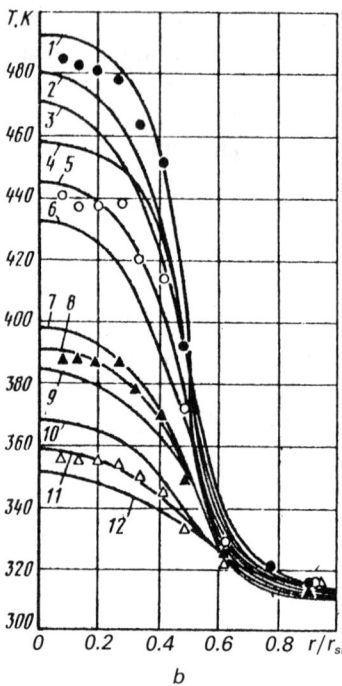

Figure 5.5 Heat carrier temperature fields in the outlet cross-section of a helical tube bundle at decreasing heat load and at $\text{Re}_b = 8.9 \cdot 10^3$ (a) and $1.75 \cdot 10^3$ (b): ●, ○, ▲, △, experimental data at $\tau = 2, 5, 9, 13$ s, respectively; 1–3, 4–6, 7–9, 10–12) design temperature fields at $\tau = 2, 5, 9, 13$ s, respectively; $1, 4$) at $K = 0.02$; $2, 5, 7, 10$) at $K = 0.045$; $3, 6, 8, 11$) at $K = 0.06$; $9, 12$) at $K = 0.10$.

the cells. Then, by analogy with (5.46), the contribution of varying an unsteady flow temperature to this resistance may be allowed for by the dimensionless parameter

$$K_{T_g}^{*'} = \frac{\partial T_f}{\partial \tau} \frac{\sqrt{\lambda_b}}{T_f} (u g \rho_b c_p)^{-1/2} = \frac{\partial T_b}{\partial \tau} \frac{1}{T_b} \sqrt{\frac{\lambda_b}{u g \rho_b c_p}} \qquad (5.49)$$

In addition, account is taken of changing turbulent flow structure in the cell of a helical tube bundle due to a wall temperature variation. As T_f, use a mean-mass flow temperature T_b. A change in the temperature T_b and its derivative $\partial T_b/\partial \tau$ is plotted in Fig. 5.6.

CALCULATING UNSTEADY HEAT AND MASS TRANSFER 137

Table 5.1 Values of dimensionless effective turbulent diffusion coefficients under unsteady conditions

Time, s	Sharp increase in heat power at Re_b			Sharp decrease in heat power at Re_b		
	$8.9 \cdot 10^3$	$1.36 \cdot 10^4$	$1.75 \cdot 10^4$	$8.9 \cdot 10^3$	$1.36 \cdot 10^4$	$1.75 \cdot 10^4$
2	—	—	—	0.025	0.0275	0.026
4	0.2	0.2	0.2	—	—	—
5	—	—	—	0.03	0.04	0.043
6	—	0.125	—	—	—	—
8	0.078	0.08	0.075	—	—	—
9	—	—	—	0.059	0.0565	0.065
10	—	0.0575	—	—	—	—
12	0.057	—	0.053	—	—	—
13	—	—	—	0.064	0.060	0.065
20	0.052	0.054	0.046	—	—	—
30	—	0.060	—	—	—	—
32	0.0585	—	0.06	—	—	—
36	—	0.0575	—	—	—	—

The dimensionless parameter K_T characterizes a contribution of unsteady heat conduction at levelling of a temperature field due to mixing of temperature nonuniformities between tube burdle cells and is expressed by the relation

$$K_T = \left(\frac{\partial T_b}{\partial \tau}\right)_{r=0} \frac{d_{sh}^2/a_{b\,eff}}{(T_{b\tau_{finite}} - T_{b\tau_0})}$$

$$= \left(\frac{\partial T_b}{\partial \tau}\right)_{r=0} \frac{d_{sh}^2 c_p \rho_b}{\lambda_{eff}(T_{b\tau_{finite}} - T_{b\tau_0})_{r=0}} \quad (5.50)$$

where $a_{b\,eff} = \lambda_{eff}/c_p\rho_b = D_t = K_{uns}ud_{eq}$, $(T_b)_{r=0}$ is the flow temperature at the centre of a helical tube bundle, i.e. at the place, at which heat release

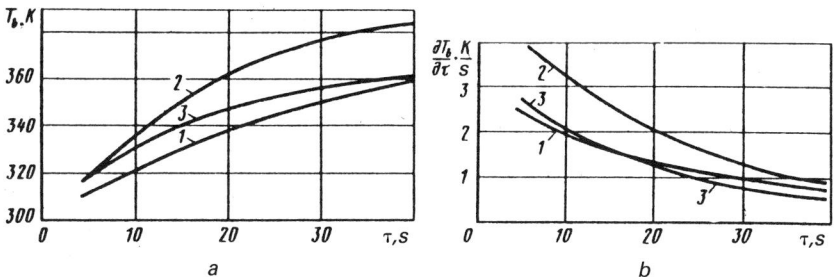

Figure 5.6 Time variations of a mean-mass temperature in a helical tube bundle (*a*) and its derivative (*b*) with increasing heat load: *1*, *2*, *3*) at Re = $8.9 \cdot 10^3$, $1.36 \cdot 10^4$, $1.75 \cdot 10^4$, respectively.

undergoes disturbance, $(T_{b\tau_{\text{finite}}} - T_{b\tau_0})$ is a temperature variation at the centre of a helical tube bundle during an unsteady process. The effective thermal conductivity, λ_{eff}, of heat carrier entering into (5.50) is unambiguously related to the coefficient D_t and the determined parameter κ. In (5.50), λ_{eff} may be, therefore, replaced by the molecular thermal conductivity λ_b. Then the parameter K_T entering into expression (5.48) is expressed by the relation

$$K_T = \left(\frac{\partial T_b}{\partial \tau}\right)_{r=0} \frac{d_{\text{sh}}^2 c_p \rho_b}{\lambda_b (T_{b\tau_{\text{finite}}} - T_{b\tau_0})_{r=0}} \tag{5.51}$$

virtually being Fourier's derivative of a dimensionless flow temperature, (5.41).

In the case of gas channel flow, the impact of unsteady heat conduction on heat transfer processes is negligible [24] because gases possess the large thermal diffusivity a_b.

The parameter K_T may be, therefore, eliminated from relation (5.48).

The dimensionless parameter

$$K_G = \frac{\partial G}{\partial \tau} \frac{d_{\text{sh}}^2}{G v_b} \tag{5.52}$$

entering into expression (5.48), defines the effect of varying a heat carrier flow rate on unsteady mixing. If in experiment a flow rate remains constant, then K_G is eliminated from (5.48).

From the aforesaid it is seen that functional relation (5.48) allows, in the course of an unsteady process, for a change both in the thermal resistance between the wall and heat carrier flow in a bundle cell and in the one between the cells if account is taken of changing the first resistance by introducing the appropriate corrections for the heat transfer coefficient, entering into equations (5.1) and (5.2), using the empirical relation:

$$\frac{\text{Nu}_{\text{uns}}}{\text{Nu}_{\text{qs}}} = f\left(K_{Tg}^*, \text{Re}_b, \frac{T_w}{T_f}\right) \tag{5.53}$$

similar, in its form, to the relation [24] for a round tube and if $\partial T_w/\partial \tau$ is determined by the successive approximation methods, then functional relation (5.48) reduces to:

$$\kappa = \frac{K_{\text{uns}}}{K_{\text{qs}}} = \kappa\left(K_{Tg}^{*\prime}, \text{Re}_b, \frac{T_w}{T_b}\right) \tag{5.54}$$

The obtained data on κ with increasing heat power over the range of the investigated parameters with no effect of Re and T_w/T_b are generalized by the relation:

$$\kappa = \frac{K_{\text{uns}}}{K_{\text{qs}}} = 1 + 4 \cdot 10^{22} \, (K_{Tg}^{*\prime})^6 \tag{5.55}$$

According to the methods used to determine λ_{eff} through the measured temperature fields of heat carrier, as seen from (5.2), the obtained values of λ_{eff} are the greater, the higher are the prescribed values of the heat transfer coefficient α. Therefore, in using the quasi-stationary values of α, the thus obtained values of λ_{eff} and κ will be underestimated with a wall temperature being increased in time when the heat transfer coefficient is greater than its quasi-stationary value and will be overestimated with the one being decreased in time ($\alpha_{\text{uns}} < \alpha_{\text{qs}}$). However, as the corresponding estimates have shown, at the rates of varying a wall temperature implemented in the present experiment, the possible deviations of the unsteady heat transfer coefficient from its quasi-stationary value are essentially less as against an error of its determination. Therefore, the quasi-stationary values of the heat transfer coefficient have been adopted to solve systems of equations (5.1)–(5.5), and the experimental data on the unsteady mixing coefficient are generalized in form (5.5). This relation is compared with the experimental data (Fig. 5.8). It is seen that at $K_{Tg}^{*\prime} = 1.5 \cdot 10^{-4}$–$2 \cdot 10^{-4}$ the experimental data greatly differ from relation (5.5). In this case, a scatter of the experimental data with respect to the Re_b numbers is random in nature, exerting no influence on the coefficient $\kappa = K_{\text{uns}}/K_{\text{qs}}$. Possibly, such a situation is bound up with a requirement for a great deal of experimental data to obtain relations of type (5.53) and (5.54). On the other side, the relations of type (5.42) well describe experimental results (Fig. 5.7). The relations of type (5.42) will be, therefore, further used to generalize the experimental data in this section. The investigations have offered determination of the coefficient K_{uns} nec-

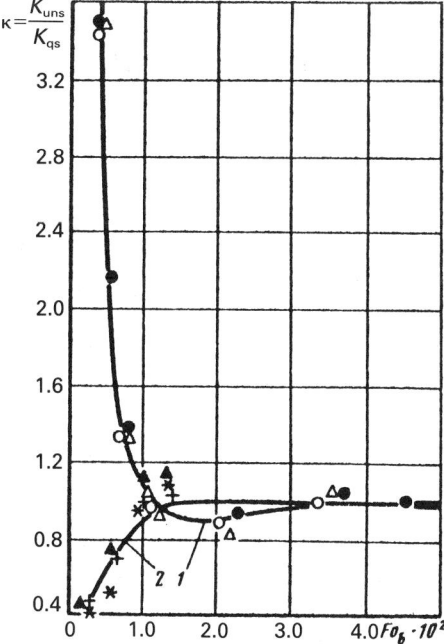

Figure 5.7 Relative unsteady turbulent diffusion coefficient vs the Fourier number: $\bigcirc$, $\bullet$, $\triangle$, experimental data for sharply increasing heat release at $Re_b = 8.9 \cdot 10^3$, $1.36 \cdot 10^4$, $1.75 \cdot 10^4$, respectively; $*$, $+$, $\blacktriangle$, experimental data for sharply decreasing heat release at $Re_b = 8.9 \cdot 10^3$, $1.36 \cdot 10^4$, $1.75 \cdot 10^4$; *1*) relation (5.43); *2*) relation (5.47).

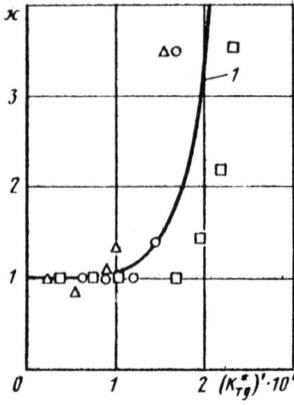

Figure 5.8 Coefficient κ vs the number $K^{*'}_{\overline{T}_g}$ at increasing heat load and at different Re_b numbers: *1*) relation (5.55); $\bigcirc$, $\square$, $\triangle$, experimental data at $Re_b = 8.9 \cdot 10^3$, $1.36 \cdot 10^4$, $1.75 \cdot 10^4$, respectively.

essary to close the system of equations (5.17)–(5.21) and to establish some new laws.

So, additional enhancement of the levelling of heat carrier temperature field nonuniformity initiated by a nonuniform heat release field due to a sharp increase in heat release and due to constant heat carrier flow rate favours the normal operation of helical tube bundles. An observed reduction in the rate of transfer processes due to a sharp decrease in heat release must be allowed for in considering transient conditions and in stopping a heat exchanger since in this case local superheatings of a tube wall are possible.

5.3 CONTRIBUTION OF THE RE NUMBER AND RATE OF ACHIEVING AN OPERATING REGIME TO UNSTEADY HEAT AND MASS TRANSFER

In Section 5.2 it has been emphasized that for the first 10 seconds, after heat load is varied sharply, the unsteady coefficient K_{uns} may substantially differ from its quasi-stationary value determined by formulas (4.15) and (4.16). In this case, relation (5.43) has been obtained over the range $Re = 8.9 \cdot 10^3 - 1.75 \cdot 10^4$ when the effect of the Reynolds number neither on the coefficient K_{qs} nor on the coefficient K_{uns} is almost observed, and in all the experiments a heating rate has been approximately the same and rather high. At the same time during operation of helical tube heat exchangers the cases of a smooth time variation of heat load may be also implemented. Moreover, it was necessary to study an unsteady heat and mass transfer process over the range $Re = (3.5-8.8) \cdot 10^3$ where the Re number affected the coefficient K_{qs} found by formula (4.16).

Experimental study of unsteady heat and mass transfer, the results of which are discussed in the present section, was made on the same set-up and by the same method, with heat power achieving a steady condition determined by the derivative $(\partial N/\partial \tau)_{max} = 0.615$–$1.1$ kW/s at time delays $\tau_0 = 3$–6 s. (In Section 5.2 the quantity $\tau_0 = 1$–1.5 s and $(\partial N/\partial \tau)_{max} = 3.64$–$7.2$ kW/s). The power regulator which exponentially varied the output generator power at increasing heat load:

$$N = N_1 + (N_2 - N_1)\left[1 - \exp\left(-\frac{\tau}{\tau_p}\right)\right] \tag{5.56}$$

where N_1 is the power before the start of a transient process, N_2 is the power at the end of a transient process and τ_p is the process time constant, was used to realize necessary levels of electric power supplied to the heated part of a bundle and prescribed with a constant time of varying the generator power. The quantity $N_1 = 0$ and N_2 was found from the condition of preserving the normal operation of a tube bundle. The quantity τ_p was chosen within the interval 1.5–15 s (a time constant of the power generator was equal to 1.5 s). The power regulator enabled one not only to program changes in the generator output power through the prescribed values of N_1, N_2 and τ_p and to initiate the generator power output according to the chosen procedure but also to stabilize the generator output power at varying load impedance. Thus, use of the power regulator promoted the modelling of different rates of achieving a steady condition, which is also observed in natural heat exchangers as well as the evaluation of their effect on the course of transient processes. In this case, the generator was regulated by a prescribed computer program.

Heat carrier temperature fields were measured in the outlet bundle cross-section using a comb composed of 10 chromel-alumel 0.1 mm dia wire thermocouples mounted on the traverse gear at the bundle cell centres at the points with the coordinates $r/r_{sh} = 0.073, 0.128, 0.193, 0.265, 0.334, 0.408, 0.479, 0.624, 0.770$ and 0.916. Since in measuring a time-varying temperature extra errors arise due to the unsteadiness of a process, the inertia of thermocouples was estimated and the time constant of their thermal inertia was found depending on a diameter of thermoelectrodes. A thermocouple under unsteady conditions is not a success in attaining instantaneously an ambient temperature and its signal is recorded with some delay. So, a time constant for 0.2 mm dia thermocouples is, on the average, 0.0026–0.008 s depending on an air flow velocity, and thermocouple inertia is estimated at temperature jumps realized for 0.04–0.02 s in experiments.

In this case, a dynamic error did not exceed an absolute measuring instrumental error. The experimental set-up equipped with the developed measuring system enabled one to determine the coefficient K_{uns} with a limiting error of $\pm 25\%$ at measuring errors of temperature—1%, of voltage—0.3%, current strength—0.5% and heat carrier flow rate—1.5%. In Section 5.2 the values of the coefficient K_{uns} were determined with the same error.

Figure 5.9 illustrates time variations of experimentally measured heat

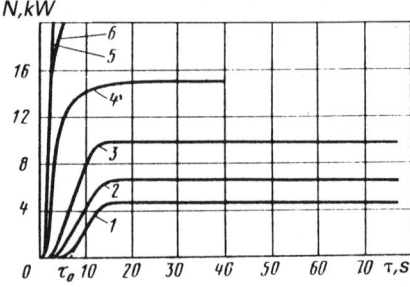

Figure 5.9 Time variation of heat power: *1–3)* at $Re_b = 3.5 \cdot 10^3$, $6.4 \cdot 10^3$, $8.8 \cdot 10^3$ and $(\partial N/\partial \tau)_{max} = 0.615, 0.727, 1.1$ kW/s, respectively; *4–6)* at $Re_b = 8.9 \cdot 10^3$ and $(\partial N/\partial \tau)_{max} = 3.64$ kW/s; $Re_b = 1.63 \cdot 10^4$, $1.75 \cdot 10^4$ and $(\partial N/\partial \tau)_{max} = 7.2$ kW/s, respectively.

release in helical tubes while Figure 5.10, time variations of a mean-mass heat carrier temperature at the bundle outlet and its derivative. As seen from Fig. 5.9, maximum rates, at which heat load achieves a steady condition, for different Re numbers strongly differ and the time, at which a rate of heating the helical tubes sharply increases, is shifted by 1–6 s from the start. This time τ_0 is equal to the quantity cut off on the τ-axis by a straight line which is determined by the maximum derivative $(\partial N/\partial \tau)_{max}$ and is a tangent to the curve $N = N(\tau)$. A maximum rate of varying a mean-mass temperature at the bundle outlet is also found by the quantity $(\partial N/\partial \tau)_{max}$ and moves up in the direction of increasing values of τ due to the thermal inertia effect.

The unsteady coefficient K_{uns} was also estimated by comparing the experimental temperature distributions at different time instants with the predicted ones as in Section 5.2. The flow model for a homogenized medium and the system of equations incorporating energy, motion, continuity and state equations as well as the heat conduction equation for a temperature distribution in helical tubes (in the bundle "skeleton") considered in Section 5.1 were used to describe unsteady flow and heat transfer processes in a helical tube bundle.

Figure 5.11 plots the predicted heat carrier temperature fields at $Re = 3.5 \cdot 10^3$ with different values of the coefficient K for different time moments $\tau = 16.8, 20.8, 24.8, 32.8, 44.8$ and 72.8 s and compares them with the

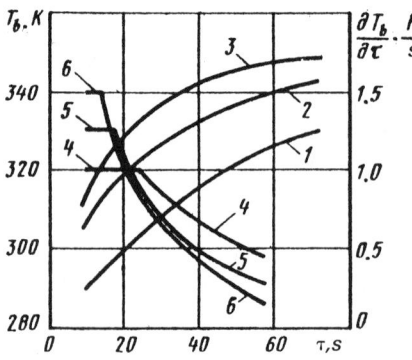

Figure 5.10 Time variations of a mean-mass heat carrier temperature and its derivative: *1–3)* T_b variation at $Re_b = 3.5 \cdot 10^3$, $6.4 \cdot 10^3$, $8.8 \cdot 10^3$, respectively; *4–6)* $\partial T_b/\partial \tau$ variation at $Re_b = 3.5 \cdot 10^3$, $6.4 \cdot 10^3$, $8.8 \cdot 10^3$, respectively.

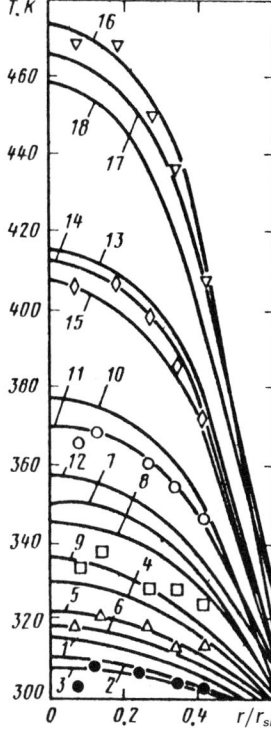

Figure 5.11 Heat carrier temperature distribution in the outlet cross-section of a helical tube bundle at $Re_b = 3.5 \cdot 10^3$: 1–18, calculation; 1–3) at $\tau = 16.8$ s and $K = 0.1, 0.2, 0.3$; 4–6) at $\tau = 20.8$ s and $K = 0.1, 0.2, 0.3$; 7–9) at $\tau = 24.8$ s and $K = 0.1, 0.2, 0.3$; 10–12) at $\tau = 32.8$ s and $K = 0.06, 0.1, 0.2$; 13–15) at $\tau = 44.8$ s and $K = 0.045, 0.06, 0.08$; 16–18) at $\tau = 72.8$ s and $K = 0.045, 0.06, 0.08$; ●, △, □, ○, ◇, ▽, experimental data obtained at $\tau = 16.8, 20.8, 24.8, 44.8, 72.8$ s, respectively.

experimentally measured temperature distributions over the range of radial coordinate $r/r_{sh} \leq 0.5$. Just this very flow region is characterized by maximum time variations of a heat carrier temperature due to sharply increasing heat power supplied to the tubes of the heated part of a bundle. A time history of a heat carrier temperature shown in Fig. 5.11 is typical of all operating conditions of a heat exchanger analyzed in the present section.

Numerical values of the coefficient K_{uns} for each time instant were determined using the modified least square method.

The thus found values of K_{uns} are listed in Table 5.2.

If the values of K_{uns} compiled in Table 5.2 are referred to its quasi-stationary value equal to $K_{qs} = 0.0585, 0.0542, 0.0521$ at $Re = 3.5 \cdot 10^3$, $6.4 \cdot 10^3$, $8.8 \cdot 10^3$, respectively, then the experimental data on $\kappa = K_{uns}/K_{qs}$ as a function of τ may be described by the relations plotted in Fig. 5.12. This plot also presents the experimental data for $Re \approx 8.9 \cdot 10^3$ taken from Section 5.2. A scatter of the experimental data almost for one and the same Re number (curves 3 and 4) shown in Fig. 5.12 may be attributed to the effect of a rate of varying heat load $(\partial N/\partial \tau)_{max}$ and to a difference in the quantity τ_0. These very reasons may be also decisive in estimating the data for other Re numbers. Indeed, comparison of the plots in Figs. 5.9 and 5.12 witnesses that a time shift in the start of a sharp increase in heat power

Table 5.2

Re_b	τ, s							
	8.8	12.8	16.8	20.8	24.8	32.8	44.8	72.8
$3.5 \cdot 10^3$	—	—	0.35	0.205	0.188	0.096	0.065	0.0575
$6.4 \cdot 10^3$	—	0.3	0.2	0.125	0.085	0.0575	0.045	0.045
$8.8 \cdot 10^3$	0.3	0.2	0.115	0.075	0.06	0.052	0.045	0.045

achieving a steady condition (Fig. 5.9) generates an appropriate time shift of the curve $\kappa = K_{uns}/K_{qs} = \kappa(\tau)$ (Fig. 5.12).

Therefore, when representing the experimental data in a dimensionless form, some conventional time allowing for the revealed effects may be included into the Fourier number instead of a real time reckoned from the starting up of a heat exchanger. This conventional time may be defined as the effective one τ_{eff}. It must take into account the time τ_0 that precedes the start of a sharp increase in heat power as well as a maximum rate, at which this heat power attains a steady condition $(\partial N/\partial \tau)_{max}$. An expression for τ_{eff} determination may be obtained from the experimental data. Then over the range of the parameters $\tau_0 = 1.5$–6 s, $(\partial N/\partial \tau)_{max} = (0.615$–$3.64)$ kW/s, $Re = 3.5 \cdot 10^3$–$8.8 \cdot 10^3$, it is possible to recommend the following formula

$$\tau_{eff} = (\tau - \tau_0)\left[a + b\left(\frac{\partial N}{\partial \tau}\right)_{max}\right] \tag{5.57}$$

to calculate the effective time taking account of τ_0 that precedes the start of sharply increasing heat power as well as the derivative $(\partial N/\partial \tau)_{max}$. Here τ is the real time reckoned from the start of operation of a heat exchanger or changes in the thermal regime of a heat exchanger, $a = 0.043$ and $b = 0.263$ s/kW. When the effective time τ_{eff} is used as a characteristic parameter, the modified Fourier number will be of the form

$$Fo_M = \frac{a\tau_{eff}}{d_{sh}^2} = \frac{\lambda_b(\tau - \tau_0)}{c_p \rho_b d_{sh}^2}\left[a + b\left(\frac{\partial N}{\partial \tau}\right)_{max}\right] \tag{5.58}$$

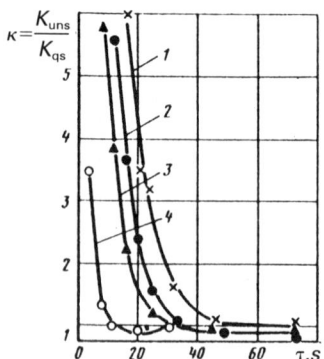

Figure 5.12 Time variation of the relative dimensionless unsteady mixing coefficient: *1–4*) experimental data at $Re_b = 3.5 \cdot 10^3$, $6.4 \cdot 10^3$, $8.8 \cdot 10^3$, $8.9 \cdot 10^3$, respectively.

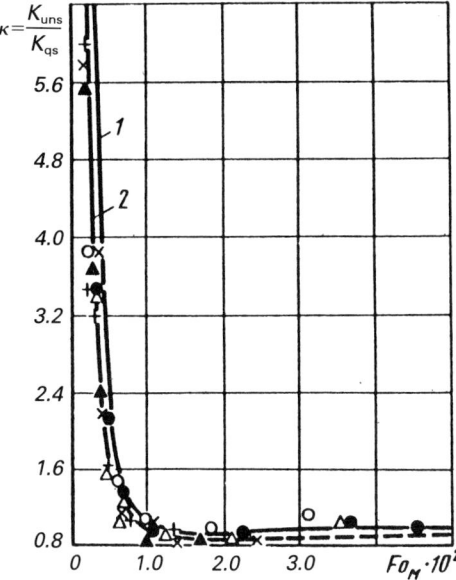

Figure 5.13 Relative dimensionless mixing coefficient as a function of the Fourier number: *1*) relation (5.43); *2*) relation (5.60); +, ▲, ×, ○, ●, △, experimental data for $Re_b = 3.5 \cdot 10^3$, $6.4 \cdot 10^3$, $8.8 \cdot 10^3$, $8.9 \cdot 10^3$, $1.36 \cdot 10^4$, $1.75 \cdot 10^4$, respectively.

In this case, the Fo_M number incorporates real physical parameters which are found from experiment. If the obtained experimental data are represented as

$$\kappa = \frac{K_{uns}}{K_{qs}} = \kappa(Fo_M) \qquad (5.59)$$

then a relative mixing coefficient over the experimental range of the parameters may be governed by the interpolation relationship

$$\kappa = 0.307 \cdot 10^{-4} Fo_M^{-2} - 0.226 \cdot 10^{-2} Fo_M^{-1} + 0.91 \qquad (5.60)$$

shown in Fig. 5.13 where it is compared with the experimental data. The value of K_{qs} is calculated by formulas (4.15) and (4.16). It should be noted that at $(\partial N/\partial \tau)_{max} \geq 3.64$ kW/s

$$\tau_{eff} = \tau - \tau_0 \qquad (5.61)$$

and

$$Fo_M = \frac{\lambda_b(\tau - \tau_0)}{c_p \rho_b d_{sh}^2} \qquad (5.62)$$

In case, the program of starting up a heat exchanger offers implementing the conditions $\tau_0 = 0$ and $(\partial N/\partial \tau)_{max} \geq 3.64$ kW/s, the effective time becomes equal to the real one from the moment of starting up a heat exchanger: $\tau_{eff} = \tau$ and $Fo_M = Fo_b$. As seen from Fig. 5.13, relation (5.60) also well strikes the experimental data for $Re_b = 1.36 \cdot 10^4$ and $1.75 \cdot 10^5$, $(\partial N/\partial \tau)_{max} = 7.2$ kW/s and $\tau = 1$ s.

Relation (5.43) may be also used to calculate K_{uns} but only for the values of $(\partial N/\partial \tau)_{max} = (3.64–7.2)$ kW/s and $\tau_0 = (1–1.5)$ s. Expression (5.43) markedly differs from the experimental data outlined in the present section (Fig. 5.13) and is finite in nature.

Thus, the studies dealing with local-averaged flow characteristics representing unsteady temperature fields of heat carrier and with the integral characteristic of the transport properties of flow K_{uns} evidence that a time variation of these characteristics at a constant heat carrier flow rate is attributed to the effect of unsteady boundary conditions at varying heat power. The observed reconstruction of temperature fields and pronounced enhancement of heat and mass transfer in a helical tube bundle at the first time moments with increasing heat power may be, as in Section 5.2., explained by altering a turbulent flow structure at unsteady heating of a bundle. Let us consider the impact of different transfer mechanisms in helical tube bundles on unsteady mixing of heat carrier: turbulent transfer, convective transfer on the cell scale and ordered transfer on the bundle diameter scale. It is know that ordered and convective transfers are affected by the Fr_M number and cannot be the original cause for heat and mass transfer augmentation at unsteady heating of a bundle. Apparently, unsteady heat transfer boundary conditions with increasing heat power promote wall layer turbulization and improvement of transfer between this layer and the flow core, i.e. wall heating increases turbulence generation in the wall layer. This process may influence an increase in vortex transfer in a bundle cell and between the cells because of convective transfer. Hence, the observed reconstruction of unsteady temperature fields of heat carrier may be, first of all, connected with enhancement of transfer of liquid portions between the wall layer and the flow core in a cell, and the ordered liquid transfer via spiral channels of helical tubes is a secondary process at unsteady heat and mass transfer. The self-similarity of the coefficient κ in terms of the Re numbers even over the range of the small numbers $Re = 3.5 \cdot 10^3$ (Fig. 5.13) may support the hypothesis about wall layer turbulization under unsteady heating of a bundle. The proposed method to generalize the experimental data on an unsteady mixing coefficient and the derived design formula may be utilized to close a system of differential equations for flow and heat transfer in such apparatuses in the homogenized flow statement and may extend the modelling potentialities of unsteady mixing processes.

5.4 UNSTEADY HEAT AND MASS TRANSFER IN HELICAL TUBE BUNDLES AT DIFFERENT Fr_M NUMBERS. GENERALIZATION OF THE EXPERIMENTAL DATA ON INCREASING HEAT LOAD

The experimental data on the unsteady effective turbulent diffusion coefficient K_{uns} introduced in Sections 5.2 and 5.3 were generalized by relation

(5.60). Relation (5.60) may be used to calculate a relative coefficient $\kappa = K_{uns}/K_{qs}$ with increasing heat load in helical tube bundles at $Fr_M = 220$ ($S/d = 12$) at $Re = 3.5 \cdot 10^3 - 1.75 \cdot 10^4$, $\tau_0 = 1-6$ s, $(\partial N/\partial \tau)_{max} = (0.615-7.2)$ kW/s. Measurement of heat carrier temperature fields in this bundle at different time moments has witnessed that the analyzed type of unsteadiness affects the coefficient κ at the first time instants due to time-varying boundary conditions at varying heat power $N = N(\tau)$. This supports the hypothesis that at unsteady heating of a bundle the turbulent flow structure is altered, thereby reconstructing temperature fields in a bundle and increasing K at the first time instants. This mechanism responsible for unsteady heat and mass transfer improvement with varying heat load will be, probably, also predominant in helical tube bundles with other Fr_M numbers. Since helical tube bundles possess the most favourable thermal and hydraulic characteristics over the range $Fr_M = 57-220$, let us consider the effect of different performance parameters on the specific features of unsteady heat and mass transfer in a helical tube bundle with $Fr_M = 57$ ($S/d = 6.1$) at increasing heat power in the same sequence as done in the case of a bundle with $Fr_M = 220$.

Unsteady temperature fields of heat carrier in a bundle with $Fr_M = 57$ were studied, and the coefficient κ was determined over the same range of $Re = 5.1 \cdot 10^3 - 1.25 \cdot 10^4$ at $(\partial N/\partial \tau)_{max} = (0.115-1.212)$ kW/s and $\tau_0 = 0-6.5$ s with heat power rapidly and slowly achieving a steady condition. Moreover, study was made of the important practical problem on the effect of a transition from one operating regime of a heat exchanger to another with a higher heat load level on the coefficient K_{uns} used to calculate unsteady temperature fields in helical tube bundles.

Unsteady heat and mass transfer in a bundle with $Fr_M = 57$ was investigated on the same set-up as in the case of a bundle with $Fr_M = 220$ when a central group of 37 helical tubes of a bundle was heated electrically. Temperature fields were measured in the outlet bundle cross-section by chromel-alumel 0.1 mm dia wire thermocouples mounted at the fixed characteristic points of the flow. The heat power differently increased in time at a constant heat carrier flow rate (Figs. 5.14 and 5.15). Figure 5.14 plots the experimental results for heat power relatively sharply achieving a steady condition (for 15 s) and for different flow rates of heat carrier, corresponding to $Re = 1.25 \cdot 10^4$, $8.9 \cdot 10^3$, $5.1 \cdot 10^3$. The relations $\kappa = \kappa(\tau)$ plotted in Fig. 5.14 were obtained by the method set forth in Sections 5.2 and 5.3 and based on comparing the experimental and predicted temperature fields of heat carrier $T = T(r, K)$ in the outlet bundle cross-section. It is seen that at the same values of a maximum heating rate $(\partial N/\partial \tau)_{max} = 1.212$ kW/s, irrespective of the Reynolds number, the coefficient κ as a function of time is described by one curve (Fig. 5.14), and a decrease in a maximum rate of heating the tubes at starting up a heat exchanger and a shift of the curve $N = N(\tau)$ from the moment of starting up a heat exchanger result in the shift of the curve $\kappa = \kappa(\tau)$ moving up in time. These results are in qualitative agreement with

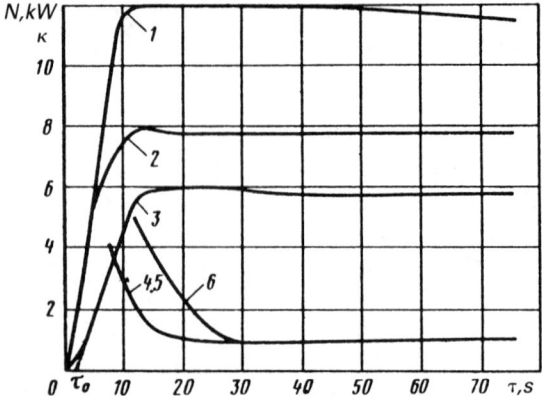

Figure 5.14 Reynolds number effect on the coefficient κ at $Fr_M = 57$: *1–3*) time variation of heat power at $Re_b = 1.25 \cdot 10^4$, $8.9 \cdot 10^3$, $5.1 \cdot 10^3$; *4–6*) time variation of κ at $Re_b = 1.25 \cdot 10^4$ and $(\partial N/\partial \tau)_{max} = 1.212$ kW/s; $Re_b = 5.1 \cdot 10^3$ and $(\partial N/\partial \tau)_{max} = 0.622$ kW/s, respectively.

the data obtained for a bundle with $Fr_M = 220$ in Sections 5.2 and 5.3 and exhibiting no effect of the Re number on $\kappa = K_{uns}/K_{qs}$ and the effect of the quantities τ_0 and $(\partial N/\partial \tau)_{max}$ on a time history of this coefficient. One can judge about the effect of the quantities $(\partial N/\partial \tau)_{max}$ and τ_0 on the coefficient K in a bundle with $Fr_M = 57$, considering Fig. 5.15 which plots the results on unsteady mixing in a helical tube bundle with $Fr_M = 57$ obtained at fixed $Re = 5.1 \cdot 10^3$ but over a wider range of the quantities τ_0 and $(\partial N/\partial \tau)_{max}$. It is seen that when heat power (curve $N = N(\tau)$) slowly achieves a steady condition at $(\partial N/\partial \tau)_{max} = 0.115$ kW/s the analyzed type of unsteadiness strongly affects the coefficient K and, hence the coefficient K_{uns} used to close a system of equations for flow in a helical tube bundle. In this case, the heat power and the coefficient κ attain their quasi-stationary values at small

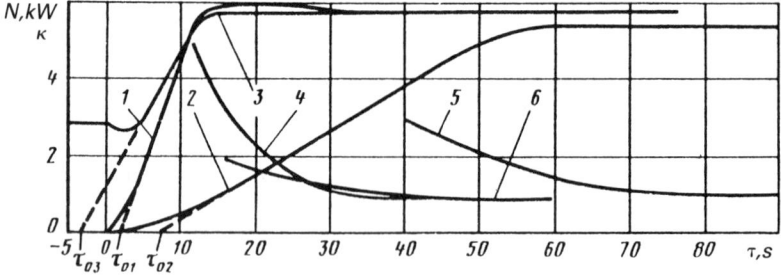

Figure 5.15 Effect of a rate of heating the helical tubes on the coefficient κ at $Fr_M = 57$ and $Re_b = 5.1 \cdot 10^3$: *1–3*) time variation of heat power at rapid, slow and step heatings at $(\partial N/\partial \tau)_{max} = 0.622$ kW/s, 0.115 kW/s, 0.35 kW/s, respectively; *4–6*) time variation of κ.

and large values of the derivative $(\partial N/\partial \tau)_{max}$ quicker than heat carrier temperatures attain their quasi-stationary values at the fixed points of the flow at the bundle outlet (Fig. 5.16).

In case of a transition from one operating condition of a helical tube heat exchanger to another with a higher heat power level (Fig. 5.15), the unsteadiness of a process also exerts an influence on the coefficient κ despite the fact that a history of temperature fields of heat carrier in this case is not so sharp (Fig. 5.17) as against the one at starting up a heat exchanger, beginning with zero heat flux (Fig. 5.16).

The results on unsteady heat and mass transfer for a bundle with $Fr_M = 57$ evidence that unsteady effective turbulent diffusion coefficient in this case is affected by the same factors as in the case of a bundle with $Fr_M = 220$. Then the experimental data on the coefficient κ for a bundle with $Fr_M = 57$ may be generalized by the same criterial relation as in the case of a bundle with $Fr_M = 220$, and formula (5.58) may be used to calculate the Fr_M number characterizing a relationship between a rate of varying a heat carrier temperature field, its physical properties and sizes of the analyzed flow region and allowing for a contribution of the quantities τ_0 and $(\partial N/\partial \tau)_{max}$ to an unsteady heat and mass transfer process in a bundle. In such data processing, the results on rapid, slow and step changes of heat power in a bundle with $Fr_M = 57$ may be generalized by one interpolation relation (Fig. 5.18):

$$\kappa = 0.114 \cdot 10^{-4}\, Fo_M^2 - 0.1053 \cdot 10^{-2}\, Fo_M^{-1} + 1.024 \qquad (5.63)$$

The derived formula points to the same mechanism of the impact of unsteady boundary conditions on a heat and mass transfer process in a helical tube bundle irrespective of the Fr_M number. Indeed, the time derivative of heat

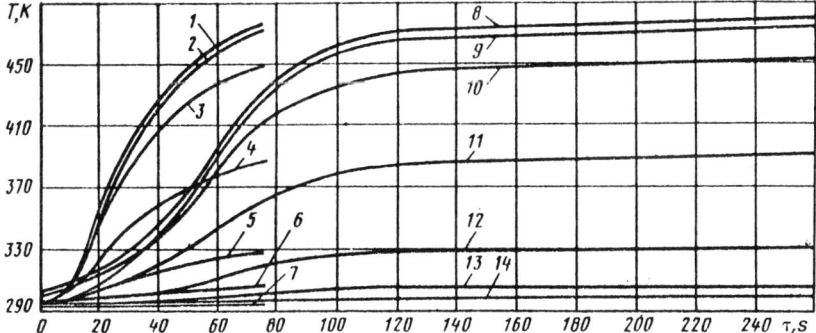

Figure 5.16 Time variation of a heat carrier temperature with heat power sharply and slowly achieving a steady condition at $Re_b = 5.1 \cdot 10^3$: *1–7)* temperature variation at heat power sharply achieving a steady condition at $(\partial N/\partial \tau)_{max} = 0.622$ kW/s for $r/r_{sh} = 0.073, 0.193, 0.334, 0.479, 0.624, 0.770, 0.916$, respectively; *8–14)* the same, at heat power slowly achieving a steady condition at $(\partial N/\partial \tau)_{max} = 0.115$ kW/s.

150 UNSTEADY HEAT AND MASS TRANSFER IN HELICAL TUBE BUNDLES

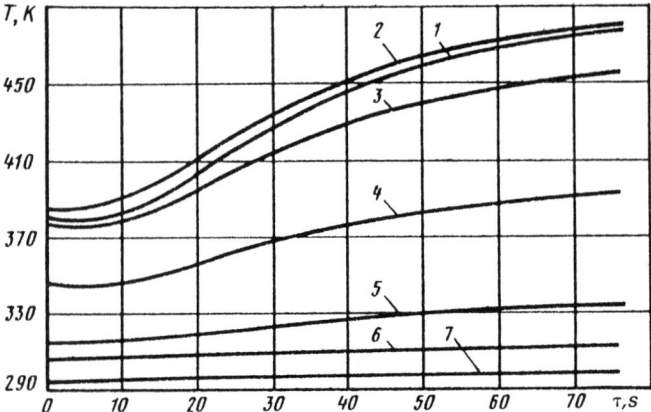

Figure 5.17 Time variation of a heat carrier temperature in going from one operating condition to a higher heat power condition at $Re_b = 5.1 \cdot 10^3$ and $(\partial N/\partial \tau)_{max} = 0.35$ kW/s: *1–7*) for $r/r_{sh} = 0.073, 0.193, 0.334, 0.479, 0.624, 0.770, 0.916$, respectively.

power $\partial N/\partial \tau$ is related to the wall temperature derivative, $\partial T_w/\partial \tau$, entering into the dimensionless parameter determined by expression (5.46) and allowing for a change in the turbulent flow structure in the wall layer with a varying tube wall temperature. Therefore, the effect of the quantity $(\partial N/\partial \tau)_{max}$ on the coefficient κ must not depend on a helical tube twisting pitch or on the Fr_M number. At the same time, with decreasing the Fr_M number (or S/d) a rate of flow swirling in a bundle increases and a flow swirling growth augments a turbulence level, first of all, in the wall layer, thus enhancing transfer processes between the wall layer and the flow core. Moreover, convective transfer between the adjacent cells of a bundle and ordered mass transfer of heat carrier via the spiral channels of the tubes in the intertube space increase. These transfer processes in a helical tube bundle must accelerate the levelling of temperature nonuniformities in the flow with decreasing the Fr_M number and in unsteady heat and mass transfer processes. Therefore, when the structure of formulas (5.63) and (5.60) is the same for bundles with $Fr_M = 57$ and 220 and the coefficient κ bears a certain identical qualitative relationship to the Fo_M number, the results calculated by (5.63) and (5.60) quantitatively differ at the same Fo_M number (Figs. 5.18 and 5.19). In this case, for a bundle with $Fr_M = 57$ the values of the coefficient κ at the first time instants are substantially less than those of κ for a bundle with $Fr_M = 220$. At $Fo_M = 10^{-2}$ the values of the coefficient κ for these bundles become virtually identical as these attain their quasi-stationary values (κ ≈ 1). Thus, in a helical tube bundle with $Fr_M = 57$ temperature nonuniformities due to unsteady heating of a bundle are levelled quicker, and a less time is needed for the coefficient κ to attain its quasi-stationary value.

If comparison is made of the values of the coefficient κ calculated by formulas (5.63) and (5.60) for $Fr_M = 57$ and 220 over the range of $Fo_M \leq 10^{-2}$ (Fig. 5.18), then $\kappa = f(Fr_M)$ may be plotted for fixed values of Fo_M (Fig. 5.19), assuming that over the range of $Fr_M = 57-220$ the coefficient κ linearly depends on the Fr_M number. With an assumption being made, the value of the coefficient κ may be found for any Fr_M number over the above range $Fr_M = 57-220$ using the relation

$$\kappa = (0.114 \cdot 10^{-4} Fo_M^{-2} - 0.1053 \cdot 10^{-2} Fo_M^{-1} + 1.024) + \Delta\kappa \tag{5.64}$$

where

$$\Delta\kappa = f(Fr_M, Fo_M) \tag{5.65}$$

Relation (5.65) may be represented as

$$\Delta\kappa = B\,(Fr_M - 57) \tag{5.66}$$

where the quantity B is determined as the tangent of the slope angle of curve (5.66) for each value of the Fo_M number (Fig. 5.19). If an interpolation formula is found for the quantity $B = B(Fo_M)$ utilizing the calculation results by relations (5.63) and (5.60) presented in Fig. 5.19, then it will be of the

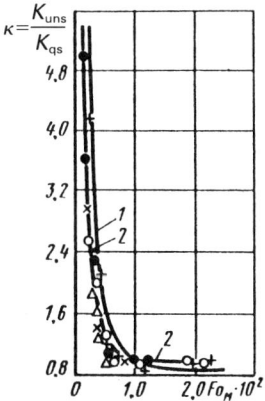

Figure 5.18 Coefficient κ vs the Fr_M number: +, ○, ●, ×, △, experimental data for $Fr = 57$ at $Re_b = 1.25 \cdot 10^4$, $(\partial N/\partial\tau)_{max} = 1.212$ kW/s, $Re_b = 8.9 \cdot 10^3$, $(\partial N/\partial\tau)_{max} = 1.212$ kW/s, $Re_b = 5.1 \cdot 10^3$, $(\partial N/\partial\tau)_{max} = 0.622$ kW/s, Re_b $5.1 \cdot 10^3$, $(\partial N/\partial\tau)_{max} = 0.115$, $Re_b = 5.1 \cdot 10^3$, $(\partial N/\partial\tau)_{max} = 0.35$ kW/s, respectively; 1) relation for $Fr_M = 220$ (5.60); 2) relation for $Fr_M = 57$ (5.63).

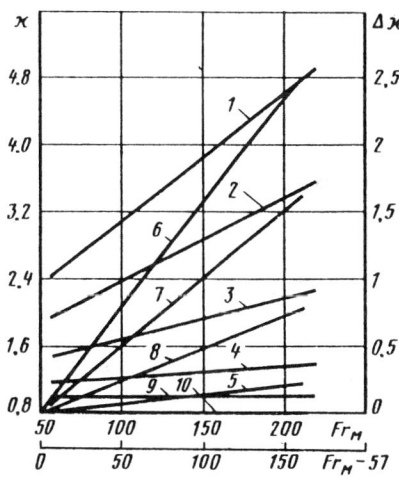

Figure 5.19 Effect of the Fr_M number on unsteady relative turbulent diffusion coefficient: 1–5) coefficient κ as a function of the Fr_M number at $Fo_M = 0.25 \cdot 10^{-2}$, $0.3 \cdot 10^{-2}$, $0.4 \cdot 10^{-2}$, $0.6 \cdot 10^{-2}$, 10^{-2}, respectively; 6–10) relation $\Delta\kappa = f(Fr_M - 57)$ at $Fo_M = 0.25 \cdot 10^{-2}$, $0.3 \cdot 10^{-2}$, $0.4 \cdot 10^{-2}$, $0.6 \cdot 10^{-2}$, 10^{-2}.

form:

$$B = (45.4 \cdot 10^6 Fo_M^2 - 18.88 \cdot 10^4 Fo_M + 245)^{-1} \quad (5.67)$$

Then relation (5.66) will be expressed as:

$$\Delta\kappa = \frac{Fr_M - 57}{45.4 \cdot 10^6 \cdot Fo_M^2 - 18.88 \cdot 10^4 Fo_M + 245} \quad (5.68)$$

and, instead of expression (5.64), we shall have:

$$\kappa = (0.114 \cdot 10^{-4} Fo_M^{-2} - 0.1053 \cdot 10^{-2} Fo_M^{-1} + 1.024)$$

$$+ \frac{Fr_M - 57}{45.4 \cdot 10^6 Fo_M^2 - 18.88 \cdot 10^4 Fo_M + 245} \quad (5.69)$$

Calculation of the coefficient κ for $Fr_M = 220$ by formula (5.69) yields a difference from the one by (5.60), not exceeding 8% over the range $Fo_M = (0.25-1) \cdot 10^{-2}$. At $Fr_M = 57$ expression (5.69) becomes identical to relation (5.63).

The above generalization of the experimental data on unsteady heat and mass transfer in helical tube bundles with $Fr_M = 57-220$ and $Fo_M = (0.25-1) \cdot 10^{-2}$ based on the physical fundamentals of a process has enabled one to assume a criterial relation which may be used to close a system of differential equations describing the flow of a homogenized medium and offering calculation of unsteady temperature fields of heat carrier and helical tubes.

The experimental results on temperature fields of heat carrier in helical tube bundles with $Fr_M = 57$ and 220 well coincide with the predicted ones estimated by solving the system of equations (5.17)–(5.21), thereby being the experimental basis for the adopted homogenized flow model, its mathematical description and calculation method.

5.5 GENERALIZATION OF THE RESULTS OF UNSTEADY HEAT AND MASS TRANSFER AT DECREASING HEAT LOAD

The experimental data on the effective diffusion coefficient K_{uns} outlined in Section 5.2 refer to a helical tube bundle with $Fr_M = 220$ and have been obtained when heat power sharply decreases from its nominal value to zero. In this case, a maximum value of the time derivative of heat power was $|\partial N/\partial \tau|_{max} = 7.5-10$ kW/s, and the revealed decrease in this coefficient, as compared to its quasi-stationary value at the first time instants, was, in nature, identical to a change in the heat transfer coefficient in round tubes for the same type of unsteadiness. In the present section of the book the earlier data are compared with the experimental results on the coefficient

K_{uns} obtained for bundles with $Fr_M = 57$ when heat power slowly achieves a steady condition, $|\partial N/\partial \tau|_{max} = 1.074-1.875$. A decrease in rates of cooling a wall (decrease in the time derivative of heat power) in this run of experiments has been provided by cooling, i.e. a transition from one operating condition of a helical tube bundle to another with less heat power (Fig. 5.20). Moreover, the operation of heat exchangers under the conditions of a transition from one operating condition to another is of independent interest. Figure 5.20 plots a time variation of heat power for a bundle with $Re = 1.25 \cdot 10^4, 8.9 \cdot 10^3, 5.1 \cdot 10^3$ as well as a temperature variation of heat carrier for $Re = 1.25 \cdot 10^4$ at the characteristic points of the flow core with the same coordinates as in the case of a helical tube bundle with $Fr_M = 220$ (Section 5.2), in a nonuniform heat release field in the bundle cross-section (electric power supply to a central group of 37 helical tubes of a bundle composed of 127 tubes). It is seen that if heating power is stabilized approximately for 16 s, then a heat carrier temperature attains a new steady level at each point of the flow practically at $\tau = 60-76$ s.

The experimentally measured temperature fields of heat carrier (Fig.

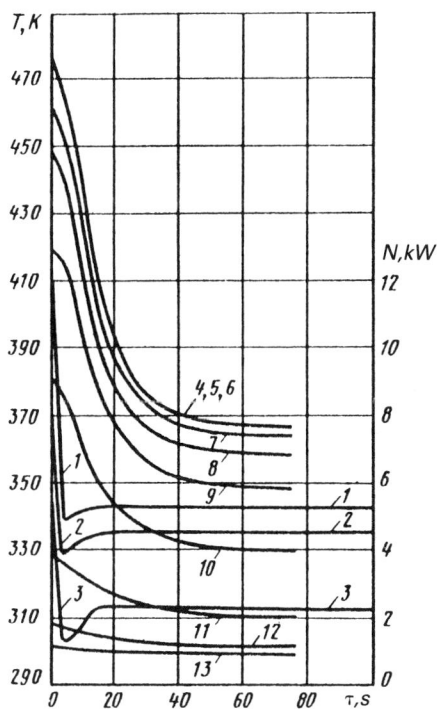

Figure 5.20 Time variations of heat power and heat carrier temperature in going on from one condition to a less heat power condition: *1–3*) heat power variation at $Re = 1.25 \cdot 10^4, 8.9 \cdot 10^3, 5.1 \cdot 10^3$, respectively; *4–13*) heat carrier temperature variation at $Re = 1.25 \cdot 10^4$ at $r/r_{sh} = 0.073, 0.128, 0.193, 0.265, 0.334, 0.408, 0.479, 0.624, 0.770, 0.916$, respectively.

5.20) are compared with the predicted ones by the method set forth in Section 5.2 (Fig. 5.21) for different time moments and coefficients K_{uns} to determine the effective diffusion coefficient K_{uns}. The temperature fields $T = T(r/r_{sh}, \tau, K)$ in Fig. 5.21 point to decreasing the coefficient K_{uns} at the first time moments, as compared to its quasi-stationary value. A time variation of the coefficient K_{uns} is shown in Fig. 5.22. It is seen that the coefficient K_{uns} virtually achieves a steady level for $\tau = 30$–40 s at $|\partial N/\partial \tau|_{max} = 1.075$–$1.875$ kW/s. At the same time for $|\partial N/\partial \tau|_{max} = 7.5$–$10$ kW/s the coefficient K_{uns} attains its quasi-stationary value for 10–13 s for a bundle with $Fr_M = 220$. Therefore, when for $|\partial N/\partial \tau|_{max} = 1.075$–$1.875$ kW/s the data are processed in form (5.42), there is a deviation from relation (5.47) (Fig. 5.23). This is attributed to the fact that the Fo_b number does not take into account a change of the wall temperature derivative $\partial T_w/\partial \tau$, to which the time derivative of heat power $\partial N/\partial \tau$ is related. In Section 5.3, the Fourier number is supplemented with the effective time τ_{eff} determined by relation (5.57)

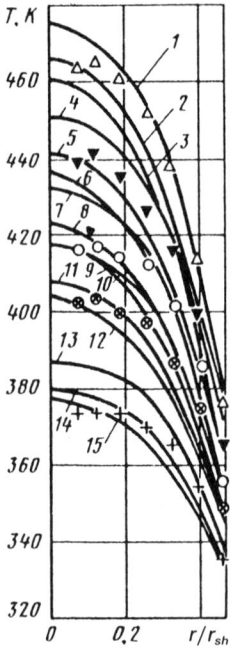

Figure 5.21 Heat carrier temperature fields in the outlet bundle cross-section at Re = $1.25 \cdot 10^4$: △, ▼, ○, ⊗, +, experimental data obtained at $\tau = 4, 8, 16, 32$ s, respectively; 1–15) design temperature fields; $1, 4, 7, 10, 13$) at $K = 0.03$; $2, 5, 8, 11, 14$) at $K = 0.06$; $3, 6, 9, 12, 15$) at $K = 0.075$.

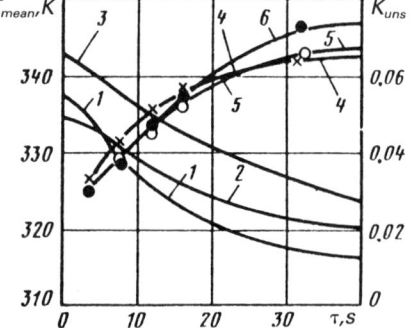

Figure 5.22 Time variations of a mean-mass temperature and coefficient K_{uns} with decreasing heat load at $Fr_M = 57$: 1–3) mean-mass temperature variation at Re = $1.25 \cdot 10^4$, $8.9 \cdot 10^3$, $5.1 \cdot 10^3$; 4–6) variation of the coefficient K_{uns} at Re = $1.25 \cdot 10^4$, $8.9 \cdot 10^3$, $5.1 \cdot 10^3$.

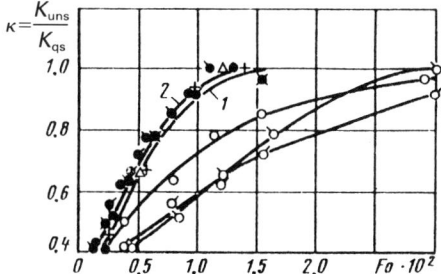

Figure 5.23 Coefficient κ vs the Fourier number at decreasing heat load: *1*) relation (5.47); *2*) relation (5.72); ●, +, △, experimental data for a bundle with $Fr_M = 220$ at $Re = 8.9 \cdot 10^3$ and $|\partial N/\partial \tau|_{max} = 7.5$ kW/s, $Re = 1.36 \cdot 10^4$ and $|\partial N/\partial \tau|_{max} = 10$ kW/s, $Re = 1.75 \cdot 10^4$ and $|\partial N/\partial \tau|_{max} = 10$ kW/s, respectively; ✖, ○, experimental data for $Fr_M = 57$, $Re = 1.25 \cdot 10^4$ and $|\partial N/\partial \tau|_{max} = 1.875$ kW/s processed using Fo_M and Fo_B, respectively; ◆, ◇, the same at $Re = 8.9 \cdot 10^3$ and $|\partial N/\partial \tau|_{max} = 1.175$ kW/s; ◐, ◓, the same at $Re = 5.1 \cdot 10^3$ and $|\partial N/\partial \tau|_{max} = 1.075$ kW/s.

instead of a real time takes into consideration a turbulent flow structure varying due to a wall temperature in the wall layer. In this case, the modified Fourier number which is calculated by formula (5.58) and depends on the quantity $|\partial N/\partial \tau|_{max}$ is used as a characteristic criterion.

For the case of sharply decreasing heat power the time τ_0 entering into expressions (5.57) and (5.58) is equal to zero. If it is assumed that for the considered type of unsteadiness the effect of the parameter $|\partial N/\partial \tau|_{max}$ on the coefficient K_{uns} is similar to that of this parameter on K_{uns} at increasing heat load, i.e.

$$\tau^{eff} = \tau\left[0.043 + 0.263\left(\frac{\partial N}{\partial \tau}\right)_{max}\right] \qquad (5.70)$$

$$Fo_M = \frac{\lambda_b \tau}{c_p \rho_b d_{sh}^2}\left[0.043 + 0.263\left(\frac{\partial N}{\partial \tau}\right)_{max}\right] \qquad (5.71)$$

then the experimental data for bundles with $|\partial N/\partial \tau|_{max} = 1.075$–$1.875$ kW/s and $Fr_M = 57$ well agree with those for $|\partial N/\partial \tau|_{max} = 7.5$–$10$ kW/s and $Fr_M = 220$ (Fig. 5.23). In this case, the experimental data on K_{uns} for bundles with $Fr_M = 57$ and 220 are referred to the quasi-stationary values of the coefficient K_{qs} obtained from experiments at each investigated operating condition in terms of the Re and Fr_M numbers. Then the experimental data on the relative coefficient $\kappa = K_{uns}/K_{qs}$ for bundles with $Fr_M = 57$ and 220 over the experimental range of Re and $|\partial N/\partial \tau|_{max}$ ($Re = 5.1 \cdot 10^3$–$1.75 \cdot 10^4$, $|\partial N/\partial \tau|_{max} = 1.075$–$10$ kW/s) may be generalized by one relation

$$\kappa = 0.454 \cdot 10^{-5} Fo_M^{-2} - 3.86 \cdot 10^{-3} Fo_M^{-1} + 1.28 \qquad (5.72)$$

valid for $Fo_M = \leq 1.4 \cdot 10^{-2}$. This relation generalizes the experimental data both for the case of heat power tending to zero (a heat exchanger ceases its operation) and for the case of decreasing heat power at a transition from one operating condition to another). The same result has been also obtained for the unsteadiness due to increasing heat power.

The Fr_M number characterizing the specific features of flow in a helical tube bundle differently affects the coefficient K_{uns} at different types of unsteadiness. If with decreasing the Fr_M number at unsteady heating of a helical tube bundle temperature nonuniformities are levelled quicker (the coefficient K_{uns} quicker attains its quasi-stationary value), then the Fr_M number does not affect κ at decreasing heat load.

This generalization of the experimental data has offered a relation to calculate an unsteady effective diffusion coefficient for operating conditions of heat exchange apparatuses and devices at heat load tending to zero as well as at a transition from one operating condition to another with less heat power. This relation may be also used to close a system of differential equations describing unsteady heat and mass transfer in helical tube bundles for the considered type of unsteadiness.

Good coincidence of the experimental and predicted temperature fields at decreasing heat power is the experimental basis for the adopted flow model, its mathematical description and calculation method and for the case of unsteady flow in helical tube bundles at decreasing heat load.

5.6 UNSTEADY HEAT AND MASS TRANSFER AT VARYING HEAT CARRIER FLOW RATE

Studies of unsteady temperature fields of heat carrier in helical tube bundles aimed at determining effective diffusion coefficients K_{uns}, when heat carrier flow rate increases and decreases, were first of all made at a rapid change of flow rate by ≈12%. In this case, investigations are, to a great extent, methodical since these trace the trends in further examination of unsteady heat and mass transfer processes for the considered type of unsteadiness being of practical importance for operation of heat exchangers. Indeed, with heat exchange apparatuses being in operation, there may exist fluctuations of a heat carrier flow rate at constant heat power as well as a transition from one operating condition to another.

Experiments were made in a 0.5 m long bundle of 127 helical tubes with $Fr_M = 57$ ($S/d = 6.1$). Electric power ($N = 8.6$ kW) was supplied to a central group composed of 37 helical tubes at increasing heat carrier flow rate from 0.125 kg/s to 0.242 kg/s, and $N = 8.2$ kW at decreasing heat carrier flow rate from 0.241 kg/s to 0.215 kg/s (Figs. 5.24 and 5.25). Thus, the Reynolds number for the analyzed type of unsteadiness varied over the range of $Re = 8.5 \cdot 10^3 - 9.6 \cdot 10^3$.

In experiment, a rapid change in heat carrier flow rate was provided by a special device outlined in Chapter 9. Figures 5.24 and 5.25 plot variations of heat carrier flow rate and temperature measured at the characteristic points of the flow with sharply increasing and decreasing air flow rate. Owing to this fact that in experiments, the parameters of starting up a heat exchanger were fixed by a 2 s time interval, beginning from the moment of varying a flow rate, it was assumed that a heat carrier flow rate attained a new value in 2 s (Figs. 5.24 and 5.25), although the used equipment operating as a camera shutter enabled a heat carrier temperature to achieve more quickly a new operating condition. In this run of experiments, a heat carrier temperature in the heated bundle zone attained its stationary value in 35–40 s after air flow rate fluctuations were made at increasing flow rate and in 25–30 s at decreasing heat carrier flow rate (Fig. 5.25).

The same method of comparing experimental and predicted temperature fields cited in Section 5.2 was adopted to determine the effective diffusion coefficient K_{uns} for this type of unsteadiness.

Heat carrier temperature fields at time-varying flow rate were calculated

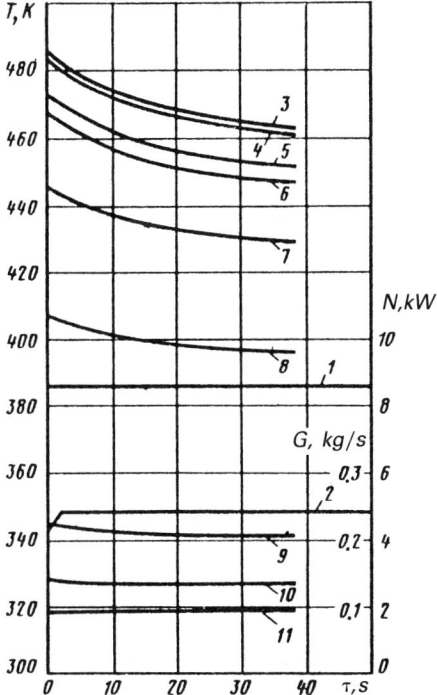

Figure 5.24 Variations of heat carrier flow rate, heat power and heat carrier temperature at flow acceleration: *1, 2*) variation of heat power and heat carrier flow rate, respectively; *3–11*) variation of heat carrier temperature for r/r_{sh} = 0.073, 0.128, 0.193, 0.265, 0.334, 0.479, 0.624, 0.770, 0.916.

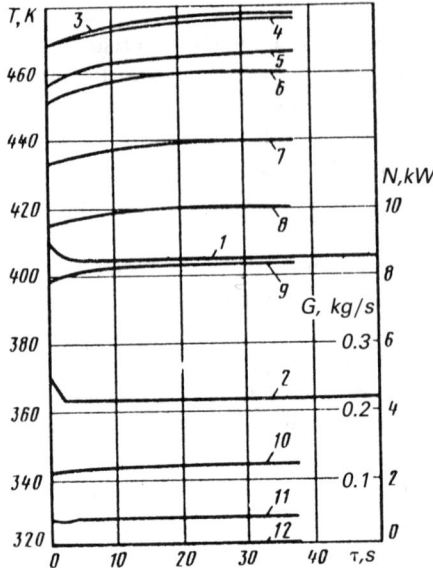

Figure 5.25 Variations of heat carrier flow rate, heat power and heat carrier temperature at flow deceleration: *1, 2*) variation of heat power and heat carrier flow rate; *3–12*) variation of heat carrier temperature for r/r_{sh} = 0.073, 0.128, 0.193, 0.265, 0.334, 0.408, 0.479, 0.624, 0.770, 0.916.

by solving numerically the system of equations (5.17)–(5.21), with the gas-dynamic equations being written to a quasi-stationary approximation using the function $G = G(\tau)$ found from experiment (Section 5.1). Figures 5.26 and 5.27 compare predicted and experimental temperature fields at certain time instants. It is seen that for the considered operating conditions of a helical tube bundle with $Fr_M = 57$ the effective diffusion coefficient K_{uns} during the whole unsteady process bound up with heat carrier flow rate fluctuations practically remains constant and equal to K_{qs} before these fluctuations are made, i.e.

$$K_{uns} = K_{qs} = 0.09 \tag{5.73}$$

The observed result may be attributed either to the fact that gasdynamic disturbances in the flow are levelled rather quickly and therefore, in this run of experiments, in which the first recording of the parameters was made in 2 s, after the air flow rate was measured, the disturbance influence on the coefficient K_{uns} could not be discovered, or to the superposition of different factors with opposite effects on the unsteady heat and mass transfer process, which yielded relation (5.73).

The thing is that in the analyzed experiments it is impossible to distinguish clearly the effect of a variable heat carrier flow rate on the coefficient

CALCULATING UNSTEADY HEAT AND MASS TRANSFER **159**

K_{uns}, i.e. the impact of flow acceleration at increasing G or that of flow deceleration at decreasing G as at varying flow rates the conditions of cooling of a helical tube wall are altered, and the derivatives $\partial T_w/\partial \tau$ and $\partial T_b/\partial \tau$ appear. Then in accord with Section 5.2 the unsteady mixing of heat carrier in a helical tube bundle must be affected by the parameters (determined by formulas (5.49) and (5.52)) which are of the form:

$$K_{Tg}^{*\prime} = \frac{\partial T_b}{\partial \tau} \frac{1}{T_b} \sqrt{\frac{\lambda_b}{ug\rho_b c_p}} \qquad (5.74)$$

$$K_G = \frac{\partial G}{\partial \tau} \frac{d_{sh}^2}{G\nu_b} \qquad (5.75)$$

The parameter K_{Tg}^* allows for the effect of an unsteady flow temperature variation on the thermal resistance between the wall and flow in a flow cell and between the cells. The parameter K_G is responsible for the effect of varying heat carrier flow rate on unsteady mixing.

It is known that in round tubes at gas heating the flow acceleration augments the quantity $K_\alpha = \mathrm{Nu}_{uns}/N_{qs}$ and the flow deceleration decreases it.

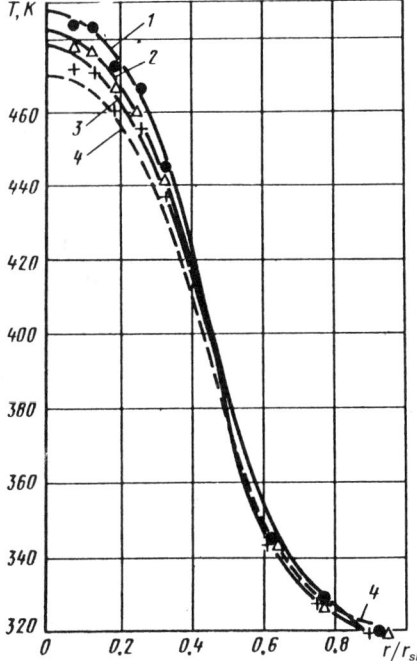

Figure 5.26 Heat carrier temperature fields in the outlet cross-section of a bundle with $\mathrm{Fr_M} = 57$ at flow acceleration: *1–3*) design temperature fields at $K = 0.09$ for $\tau = 0, 6, 12$ s, respectively; *4*) the same at $K = 0.10$ for $\tau = 12$ s; ●, △, +, experimental data obtained at $\tau = 0, 6, 12$ s, respectively.

160 UNSTEADY HEAT AND MASS TRANSFER IN HELICAL TUBE BUNDLES

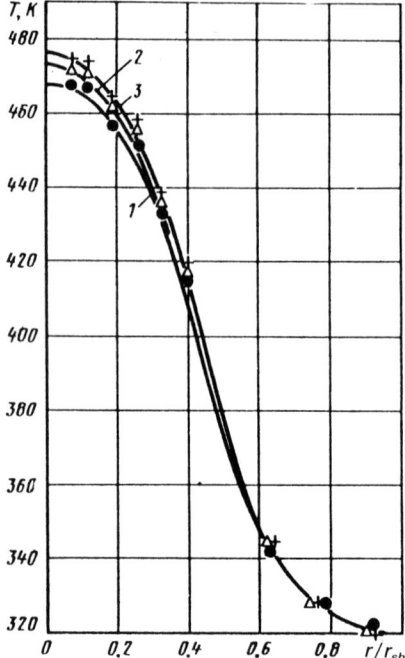

Figure 5.27 Heat carrier temperature fields in the outlet cross-section of a bundle with $\text{Fr}_M = 57$ at flow deceleration: *1–3*) design temperature fields at $K = 0.09$ for $\tau = 0, 6, 12$ s, respectively; ●, △, +, experimental data obtained at $\tau = 0, 6, 12$ s, respectively.

An assumption may be made that similar laws must manifest themselves at unsteady mixing in a helical tube bundle, i.e. the flow acceleration at gas heating must increase the coefficient $\kappa = K_{\text{uns}}/K_{\text{qs}}$ and the flow deceleration must decrease it. At the same time a wall temperature decrease at increasing G and $N = \text{const}$ must decrease the coefficient κ. Indeed, at constant heat release when a wall is cooled, extra heat is released into the flow, thereby varying a wall heat flux q_w. When a wall temperature is increased with decreasing G and $N = \text{const}$ some amount of the released heat is absorbed by the wall. This must increase the coefficient κ. Thus, the obtained relation for κ (5.73) may be a result of the effect of different parameters that specify unsteady mixing of heat carrier in a helical tube bundle according to relation (5.48).

However, the impact of these parameters for inconsiderable and considerable variations of heat carrier flow rate may be different. So, if when the heat carrier flow rate is varied by $\approx 12\%$ ($G_2/G_1 = 1.12$ and 0.89 where G_1 is the heat carrier flow rate prior to disturbances made into a system, G_2 is the new heat carrier flow rate setting in a system) the influence of an unsteady heat carrier flow temperature variation at heat carrier heating on the coefficient K_{uns} has been compensated by the opposite one on this coefficient for flow acceleration or deceleration (relation (5.73)), then the experiments made according to the above procedures, showing a heat carrier flow rate variation by $\approx 60-80\%$, have pointed to the predominant effect of an unsteady flow temperature variation on the coefficient K_{uns}. In this run of experiments, the parameters of starting up a heat exchanger were printed out with an interval 0.4 to 1.0 s. This enabled one to refine a time variation of heat carrier flow rate after the device regulating the flow rate and used as a camera shutter comes into operation.

Table 5.3 comprises time variations of the heat carrier flow rate and heat power supplied to the central group of helical tubes of a bundle for the considered operating conditions of the experimental set-up. Table 5.4 contains the values of the unsteady effective diffusion coefficients K_{uns} determined under these operating conditions.

From Table 5.4 it is seen that at the considerably increasing heat carrier flow rate ($G_2/G_1 = 1.61$, 1.77) and time-constant heat power, the coefficient K_{uns} sharply decreases at the first time instants, as compared to $K_{qs} = 0.09$ at $\tau = 0$ and smoothly increases, tending to its quasi-stationary value. Thus, for this type of the unsteadiness the coefficient K_{uns} is mainly affected not by the flow acceleration mechanism, which would have, by analogy with round tubes, increased K_{uns} or $\kappa = K_{uns}/K_{qs}$ at $T_w = $ const, but by the thermal inertia of tubes. The thermal inertia of the tubes in this case causes a time variation of heat carrier temperature fields which is identical to the one of those at decreasing heat power, at which the coefficient K_{uns} decreases at the first time instants as against the coefficient K_{qs}.

When the heat carrier flow rate ($G_2/G_1 = 0.61$, 0.665) and $N = $ const considerably decrease, the coefficient K_{uns} sharply increases at the first time

Table 5.3 Time variation of heat carrier flow rate and heat power

Ratio G_2/G_1	Parameter	Time τ, s					
		0	0.4	1	5	20	40
1.77	N, kW	6.4	6.4	6.4	6.4	6.4	6.4
	G, kg/s	0.1491	0.264	0.264	0.264	0.264	0.264
1.62	N, kW	7.5	7.5	7.5	7.5	7.5	7.5
	G, kg/s	0.1644	—	0.2666	0.2666	0.2666	0.2666
0.665	N, kW	5.7	5.7	5.7	5.7	5.7	5.7
	G, kg/s	0.2734	0.1819	0.1819	0.1819	0.1819	0.1819
0.61	N, kW	7.3	7.3	7.3	7.3	7.3	7.3
	G, kg/s	0.2525	—	0.1539	0.1539	0.1539	0.1539

Table 5.4 Values of the coefficient K_{uns} at different time moments and

Ratio G_2/G_1	Time τ,				
	0	0.8	1	2	2.8
1.77	0.09	0.0455	—	0.057	0.065
1.62	0.09	—	0.070	0.073	0.078*
1.12	0.09	—	—	0.09	—
0.89	0.09	—	—	0.09	—
0.665	0.09	0.143	—	0.120	—
0.61	0.09	0.025	—	0.23	—

*at $\tau = 3$ s, **at $\tau = 8$ s.

instants and then smoothly decreases and tends to its quasi-stationary value of K_{qs}. Hence, also, for this type of unsteadiness the coefficient K_{uns} is mainly affected not by flow deceleration due to decreasing heat carrier flow rate, which would have been observed in experiments with $T_w = $ const by analogy with round tubes but by the thermal enertia of tubes which causes the same time variation of heat carrier temperature fields as in the case of increasing heat power at constant heat carrier flow rate.

Figure 5.28 plots the coefficient K_{uns} vs a time history at increasing and decreasing relative heat carrier flow rate (G_2/G_1). When the ratio G_2/G_1 is

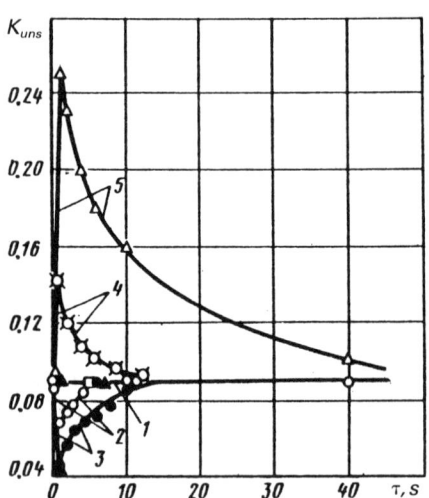

Figure 5.28 Effect of a relative change of heat carrier flow rate on the coefficient K_{uns} as a function of time: ●, ○, □, experimental data at $G_2/G_1 = 1.77, 1.62, 1.12$, respectively; △, ⌧, ▲, the same at $G_2/G_1 = 0.61, 0.665, 0.89$; 1) time history of the coefficient K_{uns} at $G_2/G_1 = 0.89$ and 1.12; 2–5) the same at $G_2/G_1 = 1.62, 1.77, 0.665, 0.61$, respectively.

Ratios G_2/G_1

s							
3.6	4	6	8.4	10	12	40
—	0.069	0.073	0.078**	0.0865	—	—
—	0.0855	0.088	—	0.09	—	0.091
—	—	0.09	—	—	0.09	—
—	—	0.09	—	—	0.09	—
0.108	—	0.103	0.097	—	0.093	—
—	0.2	0.18	—	0.16	—	0.10

small, relation (5.73) is valid both at increasing and decreasing heat power. When the ratio G_2/G_1 and heat carrier flow rate at $N = $ const are increased, the impact of this type of unsteadiness on K_{uns} and temperature field grows. So, at $G_2/G_1 = 1.62$ a minimum value of the relative coefficient $\kappa_{min} = K_{uns}/K_{qs} \approx 0.75$ and at $G_2/G_1 = 1.77$ $\kappa_{min} \approx 0.55$. When the ratio G_2/G_1 decreases up to 0.61, a maximum value of κ_{max} increases up to ≈ 2.1 (Fig. 5.28).

The revealed effects of the considered types of unsteadiness on heat carrier mixing in helical tube bundles are favourable from the point of view of the efficiency of heat exchange apparatuses and helical tube devices. So, when heat carrier flow rate at $N = $ const drastically decreases, which is possible in emergency situations, in which pipelines are broken and the amount of heat carrier is lost, the coefficient κ increases, i.e. the processes of heat carrier mixing and temperature field nonuniformity levelling are enhanced, thus facilitating thermal operating conditions of an apparatus. If the heat carrier flow rate at $N = $ const increases, then a decrease in the coefficient κ and heat carrier mixing deterioration at the first time instants do not affect the efficiency of a heat exchanger because of a pronounced decrease in a mean-mass temperature of heat carrier.

CHAPTER
SIX

METHODS OF EXPERIMENTAL STUDY OF UNSTEADY HEAT TRANSFER IN HELICAL TUBE BUNDLES

6.1 METHODS OF EXPERIMENTAL STUDY OF UNSTEADY HEAT TRANSFER

The methods of calculating convective heat transfer in terms of heat transfer coefficient, which are widely used in engineering for solving steady-state problems as well as those of predicting temperature fields under the third-kind boundary conditions can be, without any difficulties, generalized to solve unsteady-state problems.

For experimental determination of an unsteady heat transfer coefficient

$$\alpha(x, \tau) = \frac{q_w(x, \tau)}{T_w(x, \tau) - T_f(x, \tau)} \quad (6.1)$$

it is necessary to know a time variation of a mean-mass temperature of heat carrier $T_f(x, \tau)$, wall temperature $T_w(x, \tau)$ and wall heat flux density $q_w(x, \tau)$. Direct measurement of these quantities in the majority of cases is impossible. So, indirect methods must be, therefore, adopted to determine these quantities.

Indirect Method of Determining a Wall Temperature and Heat Flux Density under Unsteady Conditions

In helical tube bundles, heat transfer occurs on the outer surface of tubes (Fig. 6.1). When heat transfer is studied on the outer surface of tubes, T_w

(x, τ) and $q_w(x, \tau)$ may be determined in terms of the inner tube wall temperature $T_f(x, \tau)$ and internal heat source density $q_v(x, \tau)$ reliably measured in experiments (assuming thermal insulation of the inner wall or known heat flux density on it) by solving an inverse heat conduction problem [24, 26].

For a tube this problem is formulated as:

$$\frac{\partial^2 T}{\partial R^2} + \frac{1}{R}\frac{\partial T}{\partial R} + \frac{r_{inner}^2}{\lambda_c} q_v(\text{Fo}) = \frac{\partial T}{\partial \text{Fo}} \tag{6.2}$$

where $R = r/r_{inner}$; $\text{Fo} = a\tau(r_w - r_{inner})^2$.

The initial condition is

$$T(x, 0) = T_0(R) \tag{6.3}$$

boundary conditions are

$$T(1, \text{Fo}) = T_{inner}(\text{Fo}); \quad q_v(1, \text{Fo}) = q_{inner}(\text{Fo}) \tag{6.4}$$

Solving this problem does not face any principal difficulties. A number of methods are reviewed in [24, 26]. At present some monographs [1, 46] are published.

Determination of T_w and q_w in terms of the measured T_{inner} and q_{inner} by solving the inverse heat conduction problem is limited on physical grounds. The main point is that outer tube wall temperature variations which are finite in nature lead to finite inner tube wall temperature variations only in a finite time interval $\text{Fo}^* = a\tau^*/(r_w - r_{inner})^2$. Fo^* depends on an initial body temperature variation, rate of varying T_w and accuracy of the device recording $T_{inner}(\text{Fo})$. For example, if a temperature difference between a body and a medium at the initial time instant is $A = 200$ K, and the devices register $\Delta T_{inner} = 0.5$ K, then $\Delta T_{inner}/A = 0.0025$. In this case, at $\text{Bi} = \alpha(r_w - r_{inner})/\lambda_w = 0.01$ the parameter $\text{Fo}^* = 0.4$ and at $\text{Bi} = 0.1$ we obtain $\text{Fo}^* = 0.15$.

Any method of determining T_w and q_w in terms of measured T_{inner} and q_{inner} does not allow one to directly obtain T_w and q_w before Fo^*. However, if it is known that over the range $0 < \text{Fo} < \text{Fo}^*$ there were no sharp variations of the heat transfer conditions on the outer tube surface (i.e. if it is clear from physics of a process, then the functions $T_w(\text{Fo})$ and $q_w(\text{Fo})$ must

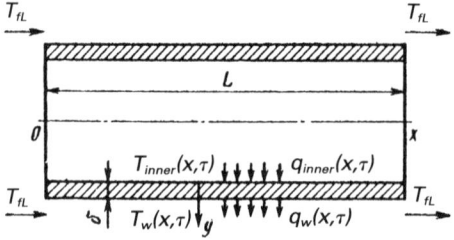

Figure 6.1 Scheme of solving the problem of external flow past a tube.

be monotonic in nature), then the relations $T_w(\text{Fo})$ and $q_w(\text{Fo})$ obtained at $\text{Fo} > \text{Fo}^*$ may be extrapolated to the range $0 \leqslant \text{Fo} < \text{Fo}^*$.

Methods of Determining a Wall Temperature and Heat Flux Density under Unsteady Conditions at Small Bi Numbers

Direct methods of determining a wall temperature and heat flux density may be substantially simplified at small values of the Bi numbers: $\text{Bi} = \alpha\delta/\lambda_w$ where α is the heat transfer coefficient, λ_w is the thermal conductivity of wall material and δ is the wall thickness. Consideration is made of the problem on unsteady heat transfer in the longitudinal heat carrier flow along the outer tube surface. It is necessary to determine an outer tube wall temperature $T_w(x, \tau)$ and heat flux density $q_w(x, \tau)$ on it in terms of the measured inner tube wall temperature $T_{\text{inner}}(x, \tau)$ and internal heat source density $q_v(x, \tau)$ at a known heat flux density (heat leakages) on the inner tube surface $q_{\text{inner}}(x, \tau)$ (Fig. 6.1).

If a tube wall thickness is small, as compared to its radius ($\delta/r < 0.2$), then the problem on heat propagation in a cylindrical wall may be reduced to the heat conduction equation for a flat plate with a reduced thickness [26]:

$$\delta_{\text{sur}} = F_w/\Pi_w \tag{6.5}$$

where F_w is the cross-sectional area of a tube and Π_w is the wetted tube perimeter. For a round tube

$$\delta_{\text{sur}} = (d_w^2 - d_{\text{inner}}^2)/4d_w \tag{6.6}$$

where d_w is the outer diameter and d_{inner} is the inner diameter.

T_w and q_w are determined under the following assumptions: (1) heat sources are located equidistant over the tube wall thickness (at a maximum wall temperature drop of ≈ 2 K the heat release nonuniformity over the thickness due to an Ohmic resistance variation does not exceed 0.2%); (2) heat leakages along the tube axis and over its perimeter are absent (i.e. the one-dimensional problem is considered; these leakages as against a radial heat flux may be neglected as a T_w variation along the bundle is close to the linear one ($\partial^2 T_w/\partial x^2 \cong 0$) and is small over the tube perimeter; (3) a temperature drop over the wall thickness $\Delta T_w = q_w \delta_{\text{sur}}/2\lambda_w$ is small, as compared to a temperature head between a wall and flow $(T_w - T_f) = q_w/\alpha$, i.e. $\text{Bi} = \alpha\delta_{\text{sur}}/\lambda_w = 2\Delta T_w(T_w - T_f)$ is small. Virtually, at $\text{Bi} < 0.1$ the heat conduction equation for a wall may be, with sufficient degree, replaced by the heat balance one, and the thermophysical properties of wall material may be considered constant for the analyzed cross-section and may be determined through the inner tube wall temperature T_{inner}. This condition is usually satisfied in the case of gas flow past thin-wall tubes. For example, in experiments with air flow in tube bundles composed of helical tubes, whose wall thickness is $\delta = 0.4$ mm over the Reynolds number range from $2 \cdot 10^3$ to $6 \cdot 10^4$ the Bi number varied within 0.0005–0.05.

Under these assumptions, the heat conduction equation assumes the form:

$$\frac{\partial T(y, \tau)}{\partial \tau} = a_w \frac{\partial^2 T(y, \tau)}{\partial y^2} + \frac{q_v(\tau)}{\rho_w c_w} \tag{6.7}$$

where y is the coordinate taken from the inner tube surface ($0 \leq y \leq \delta_{sur}$), T is the wall cross-section-variable temperature ($T_{inner} \geq T \geq T_w$ at heat carrier heating and $T_{inner} \leq T \leq T_w$ at its cooling); $a_w = \lambda_w/\rho_w/c_w$ is the thermal diffusivity of wall material; λ_w, ρ_w, c_w is the thermal conductivity, density and heat capacity of wall material, respectively, at the inner tube surface temperature T_{inner}.

Equation (6.7) is solved under the following boundary conditions:

(1) inner tube surface temperature is

at $y = 0$ $T(0, \tau) = T_{inner}(\tau)$ \hfill (6.8)

(2) inner tube surface heat flux density is

at $y = 0$ $q_{inner}(\tau) = -\lambda_w \left[\dfrac{\partial T(y, \tau)}{\partial y} \right] y = 0$ \hfill (6.9)

where $T_{inner}(\tau)$ is the measured tube wall temperature, $q_{inner}(\tau)$ are the heat leakages from the inner tube surface. If there takes place convective heat transfer on the inner tube surface, then $q_{inner}(\tau) = \alpha_{inner}(T_{inner} - T_{i.f})$ where α_{inner} is the heat transfer coefficient on the inner tube surface and $T_{i.f}$ is the mean-mass heat carrier temperature in a given tube cross-section. Usually the heat flux on the inner tube surface is much less than on the outer one (if, for example, a tube is filled with stagnant air), and it may be assumed that $q_{inner}(\tau) = 0$.

Besides boundary conditions (6.8) and (6.9) the initial condition is also prescribed

at $\tau = 0$ $T(y, 0) = T_0(y, x)$ \hfill (6.10)

i.e. a temperature distribution in a tube at the initial time moment. If at the initial time moment heat transfer is absent, then $T(y, 0) = $ const.

Equation (6.7) may be solved by the approximate method yielding the formulas convenient to calculate $T_w(\tau)$ and $q_w(\tau)$.

At low heat transfer intensity conditions when Bi $\ll 1$, a temperature profile over the tube cross-section virtually does not strain in time and varies in time in a quasi-stationary fashion. In this case, the derivative $\partial T/\partial \tau$ does not depend on the y-coordinate and may be considered to be a parameter equal to $\partial T_{inner}(\tau)/\partial \tau$, and equation (6.7) assumes the form

$$a_w \frac{d^2 T}{dy^2} = \frac{\partial T_H(\tau)}{\partial \tau} - \frac{q_v(\tau)}{\rho_w c_w} \tag{6.11}$$

METHODS OF EXPERIMENTAL STUDY 169

Solving equation (6.11) under the boundary conditions

$$q'_{inner}(x, \tau) = -\lambda_w \left[\frac{\partial T(y, \tau)}{\partial y} \right]_{y=0} \frac{\Pi_{inner}}{\Pi_w} \qquad (6.12)$$

where the inner-to-outer helical tube perimeter ratio Π_{inner}/Π_w is introduced to remove heat leakages to the outere surface, arrives at the following expression for the temperature T_w and heat flux density q_w on the outer tube surface:

$$T_w(\tau) = T_{inner}(\tau) - \frac{\delta_{sur}^2}{2a_w} \left[\frac{q_v(\tau)}{c_w \rho_w} - \frac{\partial T_{inner}(\tau)}{\partial \tau} \right] - q'_{inner}(\tau) \frac{\delta_{sur}}{\lambda_w} \qquad (6.13)$$

$$q_w(\tau) = q_v(\tau)\delta_{sur} - c_w \rho_w \delta_{sur} \frac{\partial T_{inner}(\tau)}{\partial \tau} + q'_{inner}(\tau) \qquad (6.14)$$

or with regard to (6.9) and (6.12) and $a_w = \lambda_w/\rho_w c_w$

$$T_w(\tau) = T_{inner}(\tau) - \frac{\delta_{sur}^2 q_v(\tau)}{2\lambda_w} + \frac{\delta_{sur}^2 c_w \rho_w}{2\lambda_w} \frac{\partial T_w}{\partial \tau}$$

$$- q_{inner}(\tau) \frac{\delta_{sur}}{\lambda_w} \frac{\Pi_{inner}}{\Pi_w} \qquad (6.15)$$

$$q_w(\tau) = q_v(\tau)\delta_{sur} - c_w \rho_w \delta_{sur} \frac{\partial T_{inner}(\tau)}{\partial \tau} + q_{inner}(\tau) \frac{\Pi_{inner}}{\Pi_w} \qquad (6.16)$$

These relations may be used both at heating and cooling of heat carrier flowing past the outer tube surface. In equations (6.13) and (6.14) $q_w < 0$, if heat is transferred from a heat carrier to a wall (e.g. when $q_v = 0$ and $\partial T_{inner}/\partial \tau > 0$) and, vice versa, $q_w > 0$, if heat is transferred from a wall to a heat carrier, e.g. when $(\partial_v(\tau)/c_w \rho_w) - (\partial T_{inner}(\tau)/\partial \tau) > 0$ and heat leakages from the inner tube surface are negligibly small. The heat flux density on the inner surface $q_{inner}(\tau) > 0$ if the heat flux is directed along the y-axis, i.e. when heat inflows occur on the inner surface. If as usual there occur heat leakages, then $q_{inner}(\tau) < 0$. The heat leakages $q_{inner}(\tau)$ enter into equations (6.13) and (6.14) with a "minus" sign.

This solution is well consistent with the more exact solutions of the heat conduction equation, for example, with the numerical solution obtained by the finite difference method.

In studying unsteady heat transfer in helical tube bundles heated by electric current, there will be practically no heat leakages on the inner tube surface and equations (6.13) and (6.14) reduce to:

$$T_w(x, \tau) = T_{inner}(x, \tau) - \frac{\delta_{sur}^2}{2a_w} \left[\frac{q_v(x, \tau)}{c_w \rho_w} - \frac{\partial T_{inner}(x, \tau)}{\partial \tau} \right] \qquad (6.17)$$

$$q_w(x, \tau) = q_v(x, \tau)\delta_{sur} - c_w\rho_w\delta_{sur}\frac{\partial T_{inner}(x, \tau)}{\partial \tau} \qquad (6.18)$$

As the outer tube surface temperature and heat flux density on it are calculated in terms of the measured temperature and heat leakages on the outer surface, it is necessary to estimate the time Fo*, after it has elapsed, the outer surface temperature starts varying noticeably (Fig. 6.2). This time depends on the initial body temperature distribution, rate of varying a heating surface temperature and measuring accuracy of a temperature. If it is considered that the ratio of a maximum temperature difference to a maximum temperature drop between a wall and flow (at the start or end of an unsteady process) $\delta T/(T_w - T_{inner}) = 0.01$, then at Bi = 0.004–0.08 we have Fo* = $a_w\tau^*/\delta_{sur}^2$ = 0.3–1.5 and the appropriate time for δ_{sur} = 1 mm and Kh18N10 steel wall τ^* = 0.045–0.23 s. The smaller times τ^* correspond to larger Bi, i.e. to higher heat transfer coefficients and heat carrier flow rates. The time τ^* markedly decreases for thin-wall tubes usually used in unsteady heat transfer experiments. In these cases, especially at high heat carrier flow rates, variations of inner and outer tube wall temperatures start simultaneously. For example, for δ_{sur} = 0.4 mm and Bi = 0.0003–0.05, at which the experiments were made in helical tube bundles, we obtain Fo* = 0.35–2 and τ^* = 0.017–0.097 s.

In some experiments on unsteady heat transfer in helical tube bundles which will be described below, instead of an inner tube wall surface temperature, measurement was made of a mean-mass wall temperature (Fig. 6.3)

$$T_m = \frac{\int_0^{\delta_{sur}} c_w\rho_w T dy}{\int_0^{\delta_{sur}} c_w\rho_w dy} \qquad (6.19)$$

in terms of its electric resistance.

Since at Bi ≪ 1 the derivative $\partial T/\partial \tau$ does not depend on the y-coordinates and it may be considered a parameter,

$$\left(\frac{\partial T}{\partial \tau}\right)_{y=0} = \left(\frac{\partial T}{\partial \tau}\right)_{y=\delta_{sur}} = \frac{\partial T_m}{\partial \tau} \qquad (6.20)$$

and equation (6.11) may be written as

$$a_w\frac{\partial^2 T}{\partial y^2} + \frac{q_v}{\rho_w c_w} - \frac{\partial T_m}{\partial \tau} = 0 \qquad (6.21)$$

In this case, considering a negligible temperature drop in a wall, we obtain from (6.19)

$$T_m = \frac{1}{\delta_{sur}}\int_0^{\delta_{sur}} T dy \qquad (6.22)$$

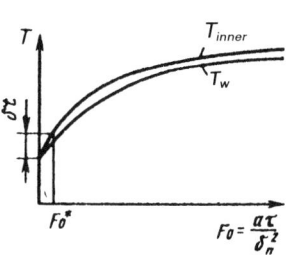
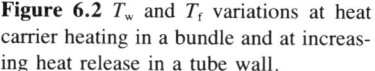
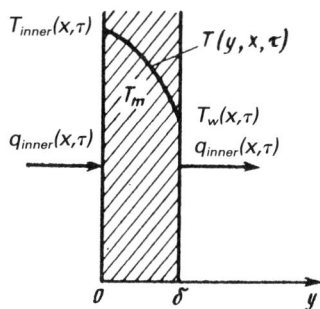

Figure 6.2 T_w and T_f variations at heat carrier heating in a bundle and at increasing heat release in a tube wall.

Figure 6.3 Scheme of solving the heat conduction problem at prescribed T_m.

On the inner tube wall the heat leakages (boundary condition (6.9)) are known. As in this case, in experiments, measurement is made of a mean-mass wall temperature, it is relatively simple (from heat balance) to find a heat flux density on the heating surface

$$q_w(\tau) = q_v(\tau)\delta_{sur} - c_w\rho_w\delta_{sur}\frac{\partial T_m(\tau)}{\partial \tau} + q_{inner}(\tau)\frac{\Pi_{inner}}{\Pi_w} \qquad (6.23)$$

This, in its turn, enables one to prescribe the second-kind boundary conditions on both surfaces to solve equation (6.21). A temperature distribution in a tube at the initial time instant (condition (6.10)) is a known initial condition necessary to solve (6.21).

A solution to equation (6.21) with regard to (6.9), (6.20) and (6.23) is of the form

$$T(y, \tau) = T_m(\tau) - \left[\frac{q_v(\tau)}{\lambda_w} - \frac{1}{a_w}\frac{\partial T_m(\tau)}{\partial \tau}\right]\left(\frac{y^2}{2} - \frac{\delta_{sur}^2}{6}\right)$$

$$- \frac{q_{inner}(\tau)}{\lambda_w}\frac{\Pi_{inner}}{\Pi_w}\left(y - \frac{\delta_{sur}}{2}\right) \qquad (6.24)$$

and a heating surface temperature is

$$T_w(\tau) = T_m(\tau) - \left[\frac{q_v(\tau)}{\lambda_w} - \frac{1}{a_w}\frac{\partial T_m(\tau)}{\partial \tau}\right]\frac{\delta_{sur}^2}{3}$$

$$- \frac{q_{inner}(\tau)}{\lambda_w}\cdot\frac{\delta_{sur}}{2}\cdot\frac{\Pi_{inner}}{\Pi_w} \qquad (6.25)$$

It is obvious that in experiments dealing with measurement of a mean-mass wall temperature, not the inverse but direct heat conduction problem is being solved and there is no need to use the restriction on the initial time Fo* which is typical of the inverse methods when a $T_{inner}(\tau)$ variation is not fixed by the devices.

Solution (6.25) is well consistent with the more exact solution to the heat conduction equation

$$\frac{\partial T(y, x, \tau)}{\partial \tau} = a_w \frac{\partial^2 T(y, x, \tau)}{\partial y^2} + \frac{q_v(x, \tau)}{c_w \rho_w} \tag{6.26}$$

obtained from assumption (6.20).

Formulas (6.25) and (6.26) are simplified if heat leakages from the inner tube surface are ignored

$$T_w(x, \tau) = T_m(x, \tau) - \frac{\delta_{sur}^2}{3} \left[\frac{q_v(x, \tau)}{\lambda_w} - \frac{1}{a_w} \frac{\partial T_m(x, \tau)}{\partial \tau} \right] \tag{6.27}$$

$$q_w(x, \tau) = q_v(x, \tau)\delta_{sur} - c_w \rho_w \delta_{sur} \frac{\partial T_m(x, \tau)}{\partial \tau} \tag{6.28}$$

It should be noted that at Bi ≪ 1 a temperature drop ($T_m - T_w$) is inconsiderable. For example, in experiments with helical tube bundles $T_m - T_w$ = 0.38 K is maximum (at $q_v = 8 \cdot 10^5$ kW/m³, $\partial T_m / \partial \tau = 100$ K/s, $\delta_{sur} =$ 0.2 mm) and, therefore, $T_w = T_m$ may be assumed.

Methods of Determining a Mean-Mass Flow Temperature Distribution along a Bundle under Unsteady Conditions

A mean-mass flow temperature $T_f(x, \tau)$ is determined in terms of the measured temperature $T_{f0}(\tau)$ at the channel inlet, mass flow rate of heat carrier $G(\tau)$ and wall heat flux density $q_w(x, \tau)$. The essence of the method to calculate $T_f(x, \tau)$ consists in solving both a one-dimensional energy equation by the method of characteristics and two Cauchy problems [26]. The energy equation is based on unit volume for one-dimensional unsteady channel flow involving heat transfer and is of the form:

$$\rho c_p \frac{\partial T_f}{\partial \tau} + \rho c_p u_f \frac{\partial T_f}{\partial x} = q_v \tag{6.29}$$

where q_v is the heat flux based on a unit volume of a flowing gas. In equation (6.29), heat supply due to dissipation is negligibly small as against q_v and is not allowed for.

In literature, objections are encountered against such writing of one-dimensional energy equation (6.29), assuming that when the differential energy equation is integrated over the channel cross-section

$$\rho c_p \frac{\partial T}{\partial \tau} + \rho c_p u \frac{\partial T}{\partial x} = -\frac{\partial q_y}{\partial y} \tag{6.30}$$

where

$$q_y = -(\lambda + \rho c_p \epsilon_q) \frac{\partial T}{\partial y} \tag{6.31}$$

the first term in LHS will contain the so-called mean-geometrical flow temperature T_g and the second, the mean-mass flow temperature T_f. It should be noted that energy equation (6.29) is not obtained by integrating the differential energy equation. The mean-mass flow temperature T_f characterizes the flow in its one-dimensional description. Inclusion of two temperatures T_f and T_g into one-dimensional equation (6.29) would exclude the possibility of their determination. For gases, usually $\partial T_f/\partial \tau \ll u_f (\partial T_f/\partial x)$ and, therefore, this problem will not arise any more. Substitution of the variables

$$X = \frac{x}{d_{eq}}; \text{Ho} = \int_0^\tau \frac{u_f(\tau)}{d_{eq}} d\tau; \Theta = \frac{T_f}{T_o} \qquad (6.32)$$

reduces equation (6.29) to the form

$$\frac{u_o}{u_f}\frac{\partial \Theta}{\partial \text{Ho}} + \frac{\partial \Theta}{\partial X} = 4\text{St}_0 \qquad (6.33)$$

Here T_0 is the characteristic temperature (e.g. $T_0 = 273$ K)

$$\text{St}_0 = f(\text{Ho}, X) = F_f \frac{q_w(\text{Ho}, X)}{T_0 c_p G(\text{Ho})} = \frac{\text{Nu}_q}{\text{RePr}}; \text{Nu}_q = \frac{q_w d_{eq}}{\lambda T_0}$$

Equation (6.29) was solved in [26] under the following assumptions: (1) c_p = const; (2) ratio of the mean velocity on the bundle section to the one in the considered cross-section $\bar{u}/u_f = 1$ (this assumption is unessential as the channel along its length is divided into several sections to determine local values of the heat transfer coefficient). The homogeneous equation, equivalent to (6.29) with regard to the above assumptions, will be

$$\frac{\partial F}{\partial \text{Ho}} + \frac{\partial F}{\partial X} + 4\text{St}_0 \frac{\partial F}{\partial \Theta} = 0 \qquad (6.34)$$

where $F(\text{Ho}, X, \Theta) = 0$. The equation of characteristics is

$$d\text{Ho} = dX = \frac{d\Theta}{4\text{St}_0 (\text{Ho}, X)} \qquad (6.35)$$

A general solution to equivalent equation (6.34) is obtained by the method of characteristics, and then the Cauchy problem is being solved, for which two regions of the course of a process must be considered.

The first region (Ho $\leq X \leq X_l = l/d_{eq}$ where l is the channel length) is valid for those channel cross-sections in which at a given moment Ho the heat carrier particles being at the moment Ho = 0 at the channel inlet have not yet reached the analyzed cross-section (unsteady heat conduction period). The unsteady heat conduction problem is being solved for this region, in which heat carrier flow heating must be considered as the solid rod one with thermal diffusivity being variable over the cross-section. In this case, a dimensionless temperature distribution along the channel at the initial dimensionless time moment Ho = 0, $\Theta = \varphi(X)$ is prescribed. A solution is of the form:

$$\Theta(\text{Ho}, X) = 4 \int_{X-\text{Ho}}^{X} \text{St}_0 (Y + \text{Ho} - X, Y) dY + \varphi(X - \text{Ho}) \quad (6.36)$$

or

$$\Theta(\text{Ho}, X) = 4 \int_{0}^{\text{Ho}} \text{St}_0 (Y, Y - \text{Ho} + X) dY + \varphi(X - \text{Ho}) \quad (6.37)$$

In studying unsteady heat transfer to gases, a gas velocity in the channel is usually no less than 10 m/s, and along the 1 m long channel a period of unsteady heat conduction even for finite channel sections does not exceed 0.1 s. Therefore, in analyzing unsteady heat transfer to gases, the first region of a process is usually not considered.

The second region ($X \ll \text{Ho} \ll \infty$), namely, the region of unsteady convective heat transfer takes place in those channel cross-sections and at those time moments, for which the initial temperature distribution is already of no importance, and a process is specified by the conditions at the channel inlet. For any cross-section of the channel this region starts when heat carrier particles at the initial time moment at the channel inlet pass through this cross-section. In this case, the flow temperature prescribed at the channel inlet: $\Theta = \varphi(\text{Ho})$ at $X = 0$ is the boundary condition. A solution to the Cauchy problem is of the form:

$$\Theta(\text{Ho}, X) = 4 \int_{0}^{X} \text{St}_0 (Y + \text{Ho} - X, Y) dY + \varphi(\text{Ho} - X) \quad (6.38)$$

or

$$\Theta(\text{Ho}, X) = 4 \int_{\text{Ho}-X}^{\text{Ho}} \text{St}_0 (Y, Y - \text{Ho} + X) dY + \varphi(\text{Ho} - X) \quad (6.39)$$

Here Y is the integration variable. If heat is transferred from a heat carrier to a wall, then q_w and q_v are negative as well as the integrals in expressions (6.38) and (6.39). In this case, the gas temperature along the channel drops.

As in studying unsteady heat transfer to gases the residence time of heat carrier in the channel usually varying from 0.003 to 0.07 s is much less than the time intervals, into which an unsteady process is divided in calculations, the heat flux in equation (6.38) may be integrated at a constant time. In this case, for each time moment account is taken of q_w as a function of coordinate alone, and equation (6.39) assumes the form

$$\Theta(\text{Ho}, X) = 4 \int_{0}^{X} \text{St}_0 (\text{Ho}, Y) dY + \varphi(\text{Ho}) \quad (6.40)$$

or in a dimensionless form

$$T_f(x, \tau) = T_{f0}(\tau) + \frac{\Pi_w}{G(\tau) c_p} \int_{0}^{x} q_w(x, \tau) dx \quad (6.41)$$

where $T_{f0}(\tau)$ is the inlet flow temperature; Π_w is the tube bundle perimeter participating in heat transfer, and $G(\tau)$ is the gas flow rate.

Thus, the expressions obtained for the heat flux density $q_w(x, \tau)$, outer tube surface temperature $T_w(x, \tau)$ and mean-mass flow temperature $T_f(x, \tau)$ enable one to determine a value of the heat transfer coefficient under unsteady conditions.

A knowledge of length and time variations of the heat flux density $q_w(x, \tau)$ offers calculating a mean-mass flow temperature at the channel outlet

$$T_{fl}(\tau) = T_{f0}(\tau) + \frac{\Pi_w}{G(\tau)c_p} \int_0^l q_w(x, \tau) dx \qquad (6.42)$$

and comparing it with the measured outlet flow temperature $T_{fl}(\tau)_{\text{measured}}$.

It should be noted that the proposed methods to determine the heat transfer coefficient under unsteady conditions makes it possible to simultaneously find the heat transfer coefficient in terms of any number of points along the channel, which considerably extends the volume of information stored in experiments, thus greatly improving the confidence and reliability of the experimental data.

6.2 DESIGN OF EXPERIMENTAL SET-UPS

The schematic of the experimental set-up is shown in Fig. 6.4. The experimental section represented a close-packed bundle of 19 helical tubes.

Helical tubes had a cross-section shaped as a circle with the cut segments and were made of steel 1Kh18N10 tubes 6 mm in dia and $\delta = 0.4$ mm in wall thickness. Two versions of helical tube bundles with twisting pitches $S = 30$ mm ($S/d = 4.15$) and $S = 90$ mm ($S/d = 12.45$) were studied. The tube bundle length between the tube plates was 855 mm, and the length of the twisted part amounted to 750 mm. A general view of the experimental section is drawn in Fig. 6.4. Bundles were heated by low-voltage current immediately conducting through them. After the bundles were assembled, the cavities in the tube plates were filled with tin to provide a contact between the tubes and the tube plates. When the bundle was heated, the tube plates were cooled with water. Air was flowing, in the longitudinal direction, past a close-packed tube bundle. High-pressure cylinder air *9* entered the experimental section via heater *10*, pressure regulator *13*, filter *12* and orifice flow meter *15*. The air flow rate was regulated by pressure regulator *13*. The air from the experimental section was ejected to atmosphere via cooling coil *25* and accessory valve *5*. A bundle was heated by direct or alternating current supplied to the tube plates via busbars *6*. Electric resistance of each tube was measured before a bundle had been assembled. A measuring circuit incorporated an accumulator, normal 0.1 Ohm resistance, semi-automatic potentiometer P 2/1. An Ohmic resistance scatter of individual tubes of a bundle did not exceed 2–3%. A maximum nonuniformity

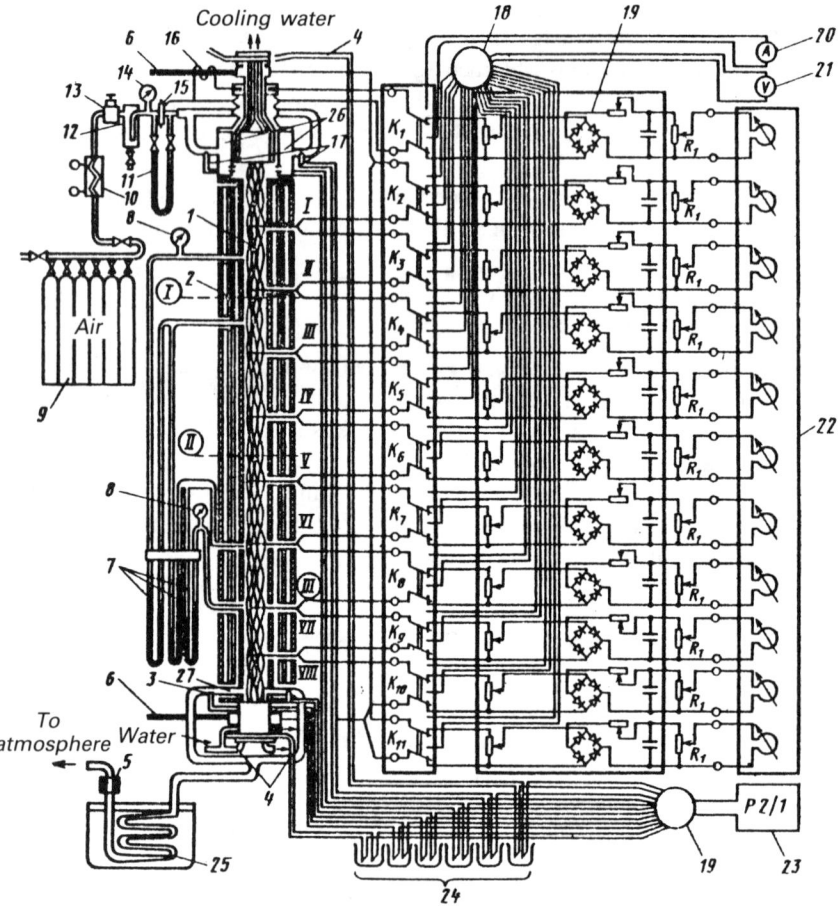

Figure 6.4 Schematic of the experimental set-up and measurements at a.c. heating of a tube bundle: *1*) tube bundle; *2*) insulation screens; *3*) thermocouples for measuring an air temperature at the experimental section outlet; *4*) leads of the wires of the thermocouples for measuring a tube wall temperature in sections *I, II, III*; *5*) valve at the experimental section outlet; *6*) busbars; *7*) differential water gauge for measuring pressure drops on the experimental section; *8*) gauges; *9*) compressed air cylinder; *10*) air heater; *11*) differential pressure gauge; *12*) filter; *13*) pressure regulator; *14*) gauge for measuring a pressure in front of an orifice; *15*) orifice flow meter; *16*) current transformer; *17*) thermocouple for measuring inlet air temperature; *18*) low-voltage current switch; *19*) rectifier; *20*) ammeter; *21*) voltmeter; *22*) oscillograph H115; *23*) semi-automatic potentiometer; *24*) Dewark flasks; *25*) cooling coil; *I–VIII*, voltage takes-off in the bundle cross-sections; *26*) upper body; *27*) outlet pipe connections.

of a current distribution over the bundle tubes due to mutual inductance amounted to 0.45%.

A tube bundle was placed into a steel hexahedral casing composed of two halves which were closely packed by poronite gaskets. The inner cavity of the casing was lined with electric insulation mica glass-ceramic plates.

The sizes of the inner cavity of the casing lined with the mica glass-ceramic plates provided close position (relative contact) of bundle tubes. The main geometrical dimensions of experimental sections are listed in Table 6.1.

The experimental sections were oriented vertically. The air entered a bundle from above (in the lateral direction) via the equidistant holes over the perimeter of the shell of the upper body (Fig. 6.4). The air was also removed via two pipe connections 27 in the lateral direction.

Two screens of sheet 0.1 mm thick steel and thermally insulating shell were located to decrease heat leakages on the outside of the body. The shell cavity was packed with asbestos. The top tube plate could move in the longitudinal direction relative to the casing to compensate a thermal expansion difference of the tube bundle and the casing. A bellows was mounted on the casing to provide air-tightness. A current-lead cylinder was arranged on the top tube plates. Electric circuit elements (busbars, tube plates, test bundle, current lead) were thoroughly insulated from the remainder parts of the experimental section.

Electric load to a tube bundle was supplied from the d.c. generator or from the autotransformer AOMK-100/0.5 and from the step-down transformer OCY-80. A maximum electric power consumed by the experimental

Table 6.1 Geometrical Dimensions of the Experimental Sections

Parameter	Notation	Dimension	Helical tubes	
			$S/d = 4.15$	$S/d = 12.45$
Diameter of a circle circumscribed around a tube cross-section	d	mm	7.24	7.23
Helical tube thickness in a tube cross-section	Δ	mm	4.51	4.15
Cross-sectional area of a tube	f_{tube}	mm²	25.9	25.15
Tube cross-section perimeter	Π_{tube}	mm	19	18.85
Flow cross-sectional area of a casing	F_{casing}	mm²	991.7	991.7
Flow cross-sectional area of a bundle	F_f	mm²	500	514
Casing cross-section perimeter	Π_{casing}	mm	116.1	116.1
Total wetted perimeter	Π	mm	477.5	476
Heated perimeter of a bundle	Π_w	mm	361	358.15
Equivalent diameter with respect to a total wetted perimeter	d_{eq}	mm	4.19	4.32
Equivalent diameter of a single central cell	$d_{eq\infty}$	mm	4.11	4.32

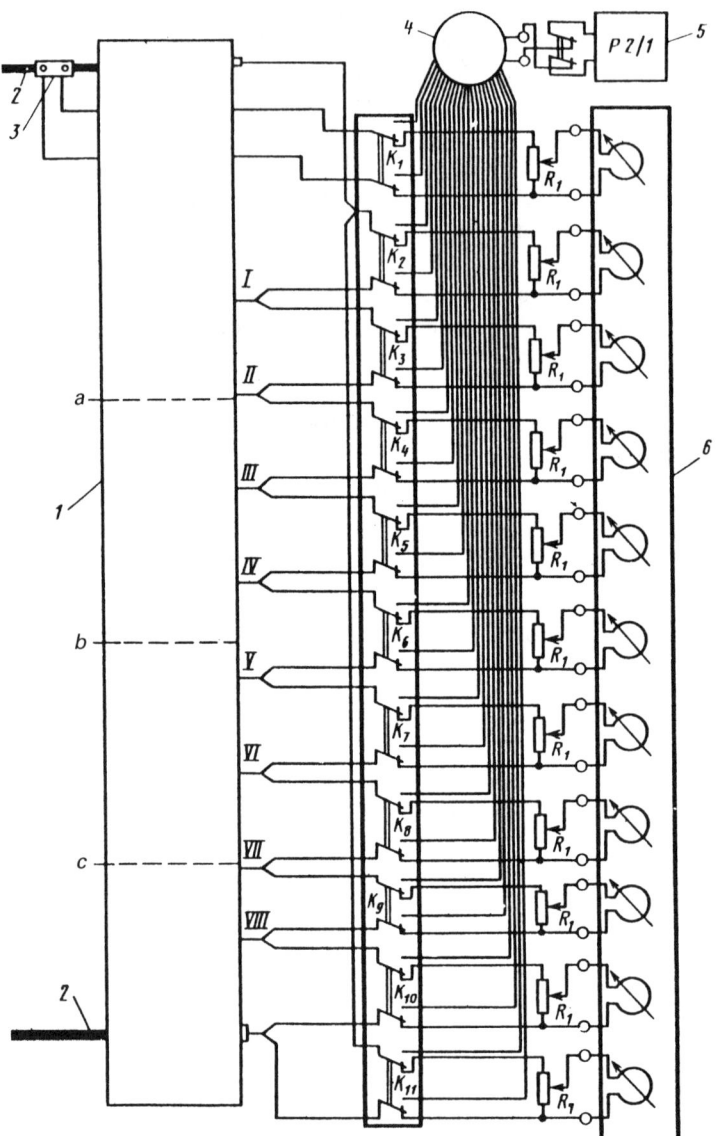

Figure 6.5 Scheme of varying a drop of voltage and strength current at d.c. heating of a bundle: *1*) experimental section; *2*) busbars; *3*) normalized resistance shunt; *4*) low-voltage current switch; *5*) semi-automatic potentiometer P 2/1; *6*) mirror-galvanometer oscillograph H115; *a*, *b*, *c*) sections in which the hot junctions of the thermocouples are located; *I–VIII*) voltage takes-off in the bundle cross-sections.

section amounted to 40 kW. Experiments were made at smooth and abrupt increases and decreases of heat load. A smooth increase in the heat load was provided by supplying the direct current of the generator being set into operation, to a tube bundle. An abrupt increase in the heat load was provided by switching on the autotransformer AOMK preliminarily tuned to a required value of the output electric power.

Hot junctions of chromel-coppel 0.2 mm dia wire thermocouples were welded into the helical tube walls in three cross-sections along the experimental sections (Fig. 6.5a, b, c) at 200, 450 and 700 mm (x/d_{eq} = 48.6, 108, 167) from the top tube plate. The thermocouples were embedded in all three cross-sections in all 19 tubes. The wires of the thermocouples provided with heat-resistant glass fibre insulation were brought out via the inner tube cavities. In unsteady processes, a temperature of the central tube of the bundle was measured by the thermocouples at 8 points at 30, 50, 100, 200, 325, 450, 575 and 700 mm (x/d_{eq} = 7.12, 13.1, 24.9, 48.6, 78.4, 108, 137, 167) from the start of heating. The hot junctions of the thermocouples were located on the broad part of the helical tube perimeter.

The specially developed methods based on electric resistance of tube material as a function of temperature were employed to measure an unsteady temperature of the tube walls. For this, the contacts capable of measuring a voltage drop on the experimental sections between these contacts were welded to the tubes in 8 cross-sections along a bundle at distances of 100, 200, 300, 396, 493, 599, 700 and 777 mm from the start of heating (x/d_{eq} = 24.4, 48.8, 73.2, 96.6, 120, 146, 171, 189). Each contact was a 0.1 mm thick and 2 mm wide two-layered steel Kh18N10 strip welded to 2–3 tubes of a bundle. The contact leads via the body were packed with teflon seals. The close-packing of the tubes in the body provided their contact. A measured voltage drop between the contacts was, therefore, applied to all the tubes of a bundle on a given experimental section.

The current strength was measured by means of resistance shunt *3* (Fig. 6.5) at d.c. heating of a bundle or by means of the transformer TT6-16 (Fig. 6.4) at a.c. heating of a bundle.

6.3 MEASURING SYSTEMS AND APPARATUSES

An experimental set-up was intended for measuring gas flow rate, flow temperature at the inlet and outlet of the experimental section, tube wall temperature in several cross-sections of a bundle, voltage drop and current strength via a tube bundle, inlet and outlet pressures and pressure drops on the experimental section under steady and unsteady conditions. The measuring systems of the set-up are shown in Figs. 6.4 and 6.5.

The orifices were used to measure gas flow rates. As these orifices mounted in small-diameter pipelines were not normalized, their gas flow rate coefficients were determined experimentally. The orifices 3.8, 13.3 and 7.6 mm dia were utilized and placed into 8, 20 and 24 ID pipelines ($m = d^2/D^2 = 0.225, 0.443, 0.103$). A rms measuring error of the averaged flow rate coefficient is 0.13%. In this case, a rms error of the gas flow rate measured by these orifices amounts to 0.62%. The limiting Reynolds number for the investigated orifices is much less than according to the norms.* A difference increases with m. It has been emphasized that gas flow rates in small-diameter pipe-lines may be rather exactly measured by the orifices although these are not normalized according to the existing rules. Moreover, the gas flow rate coefficient α and the limiting Re number differ from those recommended for the standard orifices.

Pressures were measured by the standard 0.3 class pressure gauges. Pressure drops on the orifices and on the experimental sections were measured by the mercury DT-150, DT-50 and water gauges. The chromel-alumel thermocouples specially calibrated at the All-Union Research Institute of Standards, Measures and Measuring Devices were used to measure tube wall and flow temperatures. The 0.1 class ammeters and current transformers were employed to measure the strength of current via the experimental section. A voltage drop on the experimental section was measured by a 0.1 class voltmeter. At d.c. heating, a voltage drop was measured by the potentiometer P 2/1.

Special attention was paid to measurements under unsteady conditions. The inertia of all the gauges was estimated and provided a reliable recording of the measured parameters under unsteady conditions.

A gas pressure was measured by the inductive pressure gauges DDI-21 operating together with the devices ID-211. The pressure gauges were calibrated under steady conditions by means of the standard pressure gauges and had linear characteristics. The valves placed in the pressure holes were closed to avoid gas inflows in the static pressure holes in an unsteady process. The pressure gauges at a flow temperature above 50°C were cooled with water. A delay time of the gauges did not exceed 0.001–0.005 s.

Indications of eight thermocouples embedded in the surface of the central tube of a bundle were recorded in two versions of helical tube bundles under unsteady conditions. The 0.2 mm dia wires of the thermocouples were welded directly into the tube wall and brought out via the space inside the helical tubes.

As the existing methods of calculating the inertia of thermocouples, for example, enable one to make very approximate estimates due to difficulties in finding boundary conditions on a wire surface, the appropriate experiment was made to determine the inertia of the thermocouples used. For this, a single helical tube with thermocouples being welded to its wall was mounted

---

*Rules 28-64. Liquid, gas and vapour flow rates measured by means of the standard orifices and nozzles. Moscow: Izd. Standartov, 1964. 148 p.

in a round tube and streamlined by free air. Electric current via a helical tube was abruptly switched on and off. In addition, in the first case a wall temperature increased while in the second, it decreased. The inertia of thermocouples was estimated by the almost noninertial method, when a tube wall temperature was determined in terms of its electric resistance. Later on, the thermocouple indications were supplemented with a correction $\Delta T_w = T_w - T_{w\ measured}$ (T_w is a real temperature and $T_{w\ measured}$ is a measured temperature) which depends on the wall temperature T_w and the derivative $\partial T_w/\partial \tau$. The thermocouples with the 0.1 and 0.2 mm dia wires were used to record a flow temperature under unsteady conditions. Errors due to the inertia of the thermocouples are inconsiderable, which is supported by the estimates obtained by the two-thermocouple method.

A circuit of each thermocouple is equipped with a double switch connecting it either with the potentiometer or with the oscillograph. The oscillograph records unsteady operating conditions as well as steady ones before and after an unsteady process ceases. In steady processes the thermocouples are alternately switched to the potentiometer to measure emf of the thermocouples. At this moment the galvanometer circuit in the oscillograph is disconnected to record a null of the appropriate thermocouple. As the characteristics of the galvanometers of the oscillograph are linear, the values of the measured parameters (including emf, E, of the thermocouples) are the linear functions of the deviations from the corresponding zero line on an oscillograph record. Therefore, a knowledge of initial h_1, finite h_2 and instantaneous h_i, of the deviations from the zero line and of the values of the measured parameters, for example, of the thermocouple emf $E(h_1)$ and $E(h_2)$ under steady conditions is enough to interpret oscillograph records.

Under unsteady conditions a voltage drop was also measured on the sections of a helical tube bundle. For this, the appropriate takes-off of the voltage were output to the oscillograph. The voltage drop was oscillographed on the normal resistance to measure current strength. These measurements provided determination of the electric resistance on the bundle sections. And as the electric resistance of the tube material depends on a temperature, in several experiments the wall temperature of a bundle tube was practically measured by the noninertial method. The schematic in Fig. 6.6 was adopted to find electric resistance of the tube section $l = 200$ mm as a function of temperature $R = f(t)$. A voltage drop on the tube section l on the normal resistance of emp of the thermocouples was measured by the semi-automatic potentiometer P 2/1. A system of the rheostats provided variations of the experimental tube temperature from 20 to 350° C. Switch 6 eliminated the effect of a step voltage (a voltage drop on a hot thermocouple junction) upon the thermocouple indications. The values of temperature and voltage drops on the tube section obtained in opposite directions of current to it were averaged. The relation $R = f(t)$ was determined in terms of a temperature in the middle of the section l. Similar relations were obtained in the case of

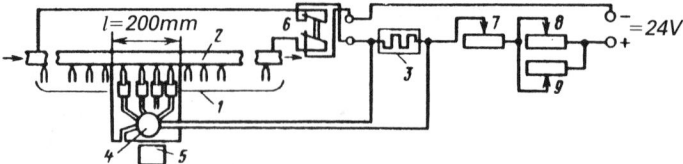

Figure 6.6 Schematic of the experimental set-up for obtaining electric resistance of a tube section vs temperature: *1*) thermocouples; *2*) tube; *3*) normal resistances; *4*) low-voltage current switch; *5*) potentiometer P 2/1; *6*) current direction switch; *7, 8, 9*) rheostats.

alternating current. A disagreement between the values of specific electric resistance over the temperature range of 50–350° C and those of steel 1Kh18N10 does not exceed 2.5%. For convenience of data processing, the relation $R = f(t)$ is reconstructed in the form $R/R_0 = \varphi(t)$ (Fig. 6.7) where R_0 is the electric resistance of the tube section at 100°C. For time variations of heat load at each chosen time instant of an unsteady process a relative value of the electric resistance R/R_0 was determined, which for simplicity of oscillograph record processing was represented as

$$\frac{R}{R_0} = \frac{R}{R_\infty} \cdot \frac{R_\infty}{R_0} \qquad (6.43)$$

where R_∞ is the electric resistance of the tube section at the stabilization moment, τ_∞, of an unsteady process. The quantity R_∞/R_0 was found as follows. A temperature in the middle of the section l was determined in terms of the wall temperature distribution along the tube at τ_∞ measured by the thermocouples. Then a value of R_∞/R_0 was found for the temperature τ_∞, using the relation $R/R_0 = \varphi(t)$. The quantity R/R_∞ may be represented in the form

$$\frac{R}{R_\infty} = \frac{U/I}{U_\infty/I_\infty} = \frac{h_U/h_{U\infty}}{h_I/h_{I\infty}} \qquad (6.44)$$

where h is the amplitude of the deviation of the oscillograph loop when the voltage drop U and current strength I are measured at an arbitrary time instant. A subscript ∞ refers to τ_∞. For the chosen time instants the values of

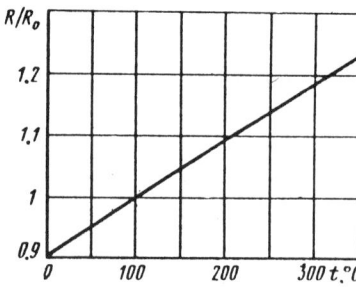

Figure 6.7 Relative electric resistance of a steel 1Kh18N10 tube wall as a function of temperature (R_0 is the resistance at 100°C).

h_U, h_I, $h_{U\infty}$ and h_I are measured on the oscillograph record, and then the quantity R/R_∞ is found. After the value of R/R_0 is determined by (6.43), the wall temperature may be found by the relation $R/R_0 = \varphi(t)$ for each time instant. Then the relation $t = f(\tau)$ is plotted.

Although an error in measuring unsteady temperatures by this method is as many as 2–3 times higher than a random error in measuring temperatures by the thermocouples, the use of this method is promising as it excludes additional errors typical of the thermocouples.

Thus, a channel wall temperature under unsteady conditions may be rather simply and practically noninertially measured by the proposed methods. An error of this method may be considerably reduced by using the oscillographs with a wider tape, by measuring the changes in the bundle section resistance by means of the bridge schemes as well as by employing the tubes, whose electric resistance of the material as a function of temperature is higher than that of steel Kh18N10. This method has enabled one to simplify the designs of experimental sections for studying unsteady processes.

The developed methods were utilized to measure unsteady temperatures of a helical tube bundle heated by alternating and direct currents (Figs. 6.4 and 6.5). For this, the drops of the voltage on each bundle section and of the current strength were oscillographed. At a.c. heating, the measuring circuit incorporated rectifier *19* (Fig. 6.4). The circuit consisted of a variable 10 kOhm resistance R_3 serving as a voltage divider, a rectifying bridge assembled of 4 diodes D7G, a variable 1 kOhm resistance R_2 necessary to initiate an output bridge signal in the region where the diode characteristics are close to linear, a 1 μΦ condenser and a variable 700–1000 Ohm resistance R_1 which provide agreement between a natural frequency of the oscillograph galvanometer and an output signal. As a result, an optimum degree of the galvanometer damping can be chosen. As the characteristics of the diodes used in the rectifier are nonlinear, a deviation of the oscillograph beam as a function of signal was determined by special calibrations for each regime over the entire range of the drop of the voltage and the current strength. An electric resistance of a bundle as a function of temperature was measured individually for nine bundle sections by the thermocouples under steady conditions.

A maximum error (corresponding to the confidence coefficient of 0.997) in measuring a mean wall temperature from 700 to 800 K amounted to 3.5 K.

6.4 EXPERIMENT PROCEDURE

Unsteady heat transfer processes due to varying heat release in the tube walls of a bundle were investigated on the earlier described experimental set-up.

184 UNSTEADY HEAT AND MASS TRANSFER IN HELICAL TUBE BUNDLES

A heat carrier temperature at the channel inlet and a heat carrier flow rate did not vary in time. Heat release in the tube bundle walls was varied abruptly or smoothly for 1–5 s.

The experiment was made in the following way: an initial steady regime was set, the values of the stationary parameters were recorded, then the measuring lines were connected with the oscillographs and the considered unsteady process took place. The parameters were recorded until a finite steady state was attained. The time of an unsteady process was specified by the direction and the degree of the disturbing impact as well as by the flow parameters at the start of a process and was varied from 5–10 s to 180–200 s. Recording at the start of a process was continuous and then discrete. After 300 s, a steady regime was recorded. All the measuring systems were calibrated at the start and at the end of each unsteady process.

Figure 6.8 displays the typical oscillograph records of unsteady processes.

6.5 DATA PROCESSING

Data processing incorporated the following stages: processing of the calibrated and stationary experiments aimed at checking the methods to measure the main parameters, oscillograph record processing and preparation of the experimental data for computer calculations. The results on the processing of the initial and finite steady regimes were compared with the existing relations. This enabled one to evaluate the applicability of an unsteady regime for further data processing.

A heat release density in tube walls was determined in terms of a measured voltage drop on them and in terms of a known electric resistance of the tube material as a function of temperature

$$q_v(x, \tau) = \bar{q}_{vw} \frac{\bar{\rho}(\bar{T}_{ms})\bar{\rho}[T_m(x, \tau)]}{[\bar{\rho}(\bar{T}_{ms})]^2} \cdot \frac{U_\tau^2}{U_s^2} \tag{6.45}$$

where

$$\bar{q}_{vw} = \frac{I_s U_s}{l_0 F_w} = \frac{I_s^2 \bar{\rho}(\bar{T}_{ms})}{F_w^2} \tag{6.46}$$

is the heat release density in a steady regime, l_0 the heated length, F_w the cross-sectional area of the tube bundle walls, U_s and U_τ, the voltage drop in the steady and unsteady regime, respectively; I_s, the current strength in a bundle in a steady regime, $\bar{\rho}$ the specific electric resistance of the tube material; $\bar{T}_{ms}$, $\bar{T}_{m\tau}$, the bundle tube wall temperature averaged over the cross-section and along the experimental section under steady and unsteady conditions, respectively. In measuring the current strength in an unsteady regime

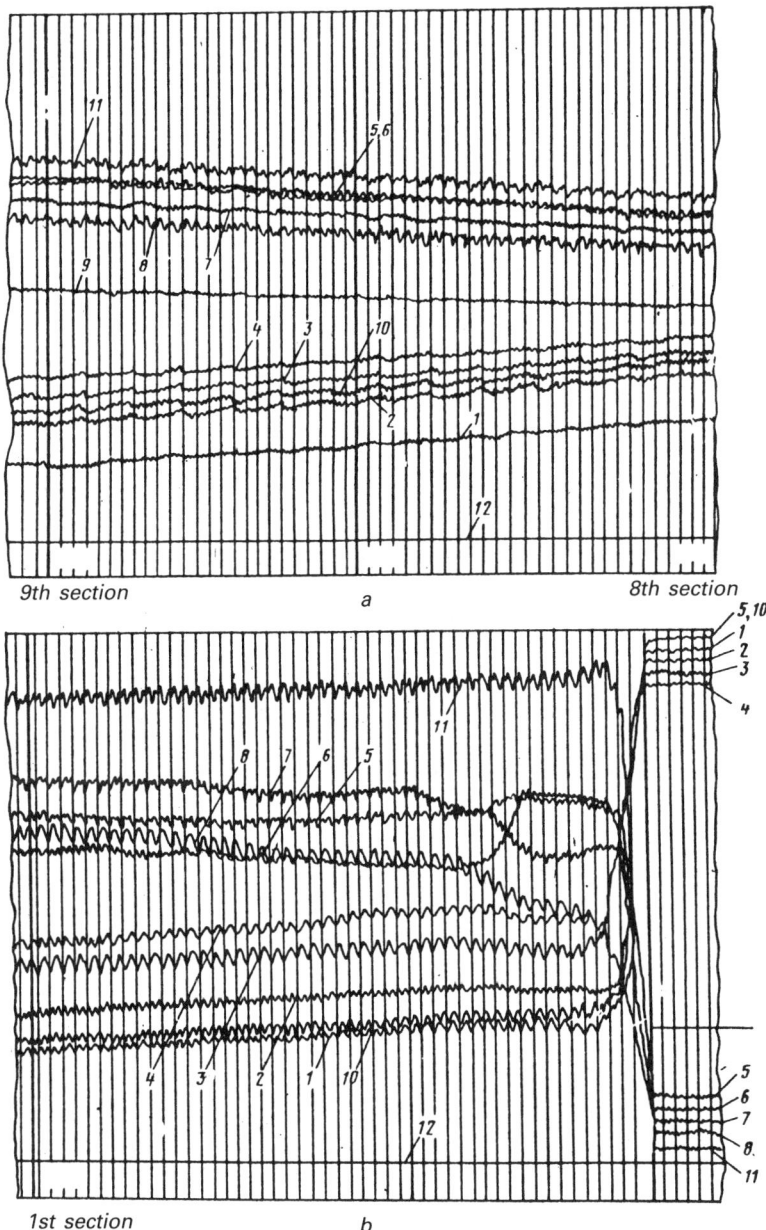

Figure 6.8 Oscillograph records of unsteady heat transfer processes: *a*) smooth increase in heat load; *b*) sharp increase in heat load; *1–9*) voltage drop on the bundle sections; *10*) total voltage drop; *11*) current strength; *12*) zero line.

$$q_v(x, \tau) = \frac{I_\tau^2 \rho [T_m(x, \tau)]}{F_w^2} \tag{6.47}$$

An unsteady heat transfer coefficient-to-its quasi-stationary value ratio was found as follows:

$$K_\alpha = \frac{Nu_f}{Nu_{f.qs}} \Big/ \frac{Nu_{f0}}{Nu_{f.qs0}} \tag{6.48}$$

where Nu_f and $Nu_{f\,qs}$ are the experimentally found Nusselt numbers under unsteady and steady conditions (initial, with decreasing heat load or with increasing flow rate; finite, with increasing heat load or with decreasing flow rate); Nu_{f0} and $Nu_{f\,qs0}$ are the Nusselt numbers calculated in terms of the relations for steady heat transfer obtained in preliminary experiments using the experimental values of Re_f and T_w/T_f under unsteady and steady conditions, respectively.

Heat balance convergence was controlled by comparing the measured $T_{fl}(\tau)_{measured}$ and calculated $T_{fl}(\tau)$ mean-mass flow temperatures at the experimental section outlet. A heat balance discrepancy determined as $[T_{fl}(\tau)_{measured} - T_{fl}(\tau)]/[T_{fl}(\tau) - T_{f0}(\tau)]$ did not exceed 10–15%.

An experimental error was estimated. In this case, it was assumed that the quantity φ was measured by the devices with random errors distributed according to the normal law. A maximum error of $\delta(\varphi)$ in determining φ was equal to a three-fold value of a rms error, i.e. the confidence coefficient of 0.997 was ascribed to the confidence interval $\pm \delta(\varphi)$.

A maximum error in measuring the mean Nusselt number under steady conditions is $\delta(\overline{Nu})_{measured} = 5.6\%$. As the experiment is aimed at obtaining the relation $Nu = f(Re)$, the found value of the Nu number must be referred to a specific value of the Re number which is also determined with some error. In this case, $\delta(Re) = 4\%$. The reference error is

$$\delta \overline{Nu}(Re)_{ref} = m\delta(\overline{Re}) = 3.2\% \tag{6.49}$$

where $Nu = CRe^m$. The reference error with respect to T_w/T_f is $\delta Nu(T_w/T_f)_{ref} = 0.7\%$. A total maximum error in determining the mean Nusselt number is

$$\delta(\overline{Nu}) = \sqrt{\delta(\overline{Nu})_{measured}^2 + \delta(\overline{Nu})_{ref}^2} = 6.5\% \tag{6.50}$$

For the local Nusselt number the similar estimates yield $\delta(Nu) = 10\%$.

In determining heat transfer coefficient under unsteady conditions, the additional errors are made, as compared to the steady ones. First, these are the errors due to time-varying quantities measured by the oscillograph. Second, in some experiments, with increasing heat load at a constant heat carrier flow rate or with decreasing it at constant heat release when a wall temperature increases, only some amount of heat released in the bundle tube walls is spent for heat carrier heating. A density of the heat flux, q_w, to a heat carrier is determined by formula (6.18) or (6.28) as a difference between the internal heat release in the tube walls and the heat spent for heating of the wall itself. An error in determining q_w is the larger, the smaller is the ratio $q_w/q_v \delta_{sur}$. At the start of the above unsteady processes the greater amount of the heat released in the wall is spent for its heating, the ratio $q_w/q_v \delta_{sur}$ is small and the error of δq_w is maximum. At the end of the unsteady processes the quantity $q_w q_v \delta_{sur}$ tends to 1 and the error of δq_w is close to that of q_w under steady conditions. At last, an additional error is attributed to the fact that the amount of the heat spent for wall heating

$$c_w \rho_w \delta_{sur} \frac{\partial T_{inner}(x, \tau)}{\partial \tau}$$

and an error in determining the derivative $\partial T_{inner}/\partial \tau$ by the numerical methods is always greater than the one associated with finding the initial relation $T_{inner}(\tau)$. In experiments, with decreasing heat load, the wall heat flux density q_w is determined by the derivative $\partial T_{inner}/\partial \tau$ (at $q_v = 0$) which is more exact, the greater is its value. Therefore, at the start of these unsteady thermal processes a measuring error is minimum. As far as a wall temperature is stabilized, it substantially increases to the end of an unsteady process.

An error in determining the wall heat flux density δq_w was calculated, assuming that $q_w/q_w \delta_{sur} = 0.5$ with increasing a wall temperature (heat load increases and gas flow rate decreases) and that $q_w/q_v \delta_{sur} = 1.5$ with decreasing a wall temperature (heat load decreases and gas flow rate increases).

According to the calculations made, at maximum errors

$\delta G = 1.8\%$, $\delta(q_v \delta_{sur}) = 2\%$, and $\delta(c_w \rho_w \delta_{sur} \partial T_{inner}/\partial \tau) = 8.6\%$

at increasing heat load $\delta q_w = 9.5\%$ and at decreasing heat load, 5.8%. In this case, at a temperature head of the order of $(T_w - T_f) = 100$ K, the measuring errors of $\delta Nu_{measured} = 12.2\%$ at increasing heat load and $\delta Nu_{measured} = 9.2\%$, and $\delta K_{\alpha measured} - 13$ and 9.5%, respectively. Considering a reference error at increasing heat load, we have $\delta K_\alpha = 18\%$ and at decreasing heat load, $\delta K_\alpha = 11\%$. As mentioned above, the confidence coefficient of 0.997 corresponds to the confidence interval $\pm \delta K_\alpha$. It should be noted that the above figures estimate errors of individual experimental points and that the errors of the obtained averaged relations are much less.

CHAPTER
SEVEN

UNSTEADY HEAT TRANSFER IN HELICAL TUBE BUNDLES

7.1 GENERAL INFORMATION

The present chapter is concerned with the results of experimental study of the unsteady heat transfer coefficient in in-line helical tube bundles obtained by the methods and on the set-ups described in Chapter 6. Naturally, these studies do not cover all the possible types of unsteady processes. Therefore, a short description of the experimental results on unsteady heat transfer in round tubes over a wide range of performance parameters in the presence of different types of unsteady impacts [24, 26] encountered in practice precedes, in the present chapter, the analysis of unsteady heat transfer in helical tube bundles. A knowledge of these results is necessary to make comparison with the helical tube bundle data as well as to qualitatively evaluate a contribution of different unsteady impacts to heat transfer if there are no direct experiments in helical tube bundles.

7.2 RESULTS OF EXPERIMENTAL STUDY OF UNSTEADY HEAT TRANSFER IN TUBES

Experimental study was made of unsteady heat transfer in gas and liquid tube flows. Experiments were performed under the following unsteady boundary conditions:
   (1) a wall temperature was varied due to an abrupt or a smooth change

of heat release in the tube wall at a constant gas flow rate and heating in a tube;

(2) a tube wall was heated at sharply and smoothly increasing hot gas temperature at the tube inlet. The gas flow rate remained constant;

(3) a flow rate of a gas heated in the tube was varied. Simultaneously, heat release was varied in the tube wall in a fashion not to cause time wall temperature changes;

(4) under gas heating and cooling conditions, a gas flow rate was varied, thereby causing wall temperature variations;

(5) experiments similar to those in Items 1 and 4 were made at tube liquid heating.

Experiments were made in 5.39–42.8 mm dia tubes over the following ranges of the parameters: for a gas, $Re_f = 6 \cdot 10^3$–$6 \cdot 10^5$, temperature factor $T_w/T_f = 0.3$–1.7, rate of varying both a wall temperature $\partial T_w/\partial \tau = -550$–$700$ K/s and gas flow rate $dG/d\tau = -0.024$–0.007 kg/s^2; for a liquid, $Re_f = 5 \cdot 10^3$–10^5, $Pr_f = 2$–12, $Pr_f/Pr_w = 1$–37, $\partial T_w/\partial \tau = -120$–$318$ K/s, $dG/d\tau = -0.4$–0.5 kg/s^2.

Over the investigated ranges of the parameters the unsteady heat transfer coefficient-to-appropriate quasi-stationary value ratio $K_\alpha = Nu/Nu_{qs}$ was varied within 0.4–3.5.

If at a constant gas flow rate, the tube wall heat release increases abruptly, then a tube wall temperature grows with a decreasing rate, asymptotically approaching a steady temperature corresponding to a prescribed heat flux (Fig. 7.1). At $\tau = $ const the wall temperature and its derivative $\partial T_w/\partial \tau$ increase with x/d. A wall temperature stabilization time increases with x/d.

At $x/d = $ const the wall temperature stabilization time is the less, the smaller is the heat load and the larger is Re_f. When the wall thickness decreases the wall temperature stabilization time drops approximately proportional to the wall thickness.

At increasing electric load the quantity $\partial T_w/\partial \tau$ attains its maximum at the initial time moment and then asymptotically drops to 0. This is consistent with decreasing K_α from its maximum to 1. For the investigated tubes $\delta = 0.165$–0.3 mm in thickness the heat transfer coefficient has attained its quasi-stationary value (K_α does not exceed 1.05) in 0.5–8 s after the electric load has been applied. The stabilization time of the heat transfer coefficient is much less than that of the wall temperature and decreases with increasing Re_f.

With lowering electric load (Fig. 7.1) the wall temperature drops with a decreasing rate, its stabilization starting the earlier, the less is x/d. The quantity K_α is varied, in this case, from 0.6–0.8 to 1. The stabilization time of the heat transfer coefficient amounts to 0.5–15 s.

The data plotted in Fig. 7.1 illustrate that unsteady operating conditions with the same mass flow rates and heat loads are characterized by almost the same values of T_w and K_α at the identical time instants irrespective of a

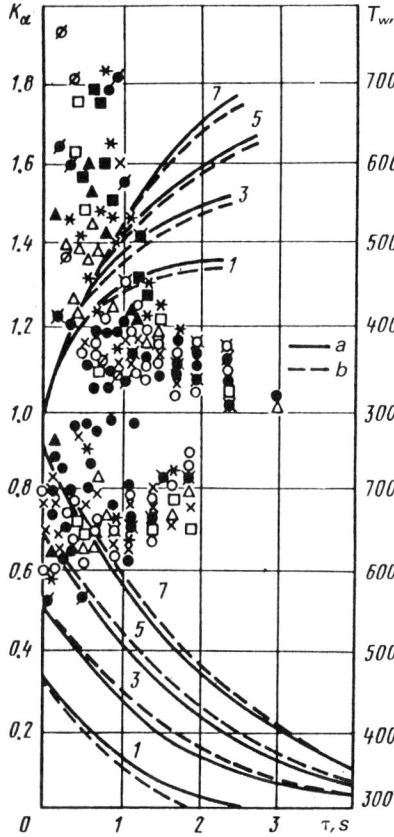

Figure 7.1 Variations of a wall temperature (curves) and K_α (points) at different x/d with increasing and decreasing heat load: heat load increase ($K_\alpha > 1$): $Re_f = (1.42-2) \cdot 10^5$, $T_w/T_f = 1-1.29$; a) $p_1 = 21.8 \cdot 10^5$ Pa; b) $p_1 = 7.56 \cdot 10^5$ Pa. Heat load decrease ($K_\alpha < 1$): $Re_f = (8.3-12.2) \cdot 10^4$, $T_w/T_f = 1-1.4$, a) $p_1 = 4.6 \cdot 10^5$ Pa; b) $p_1 = 9 \cdot 10^5$ Pa; ⌀, ⋈, ×, △, □, *, ∅, x/d = 17.1, 44, 71, 98, 126, 152, 179, respectively, for version a; ●, ⋈, ●, ▲, ■, ●, ✳, x/d for version b.

gas pressure. Thus, a pressure change does not affect an unsteady heat transfer process (as well as on a steady heat transfer process). This has enabled one to choose dimensionless parameters (1.82)–(1.85) used for experimental data generalization.

The empirical relations are obtained for practical calculations of unsteady heat transfer involving gas and liquid tube flows as well as time-varying heat release in the channel walls and heat carrier flow rate [24].

The relation for $\Delta K_{\alpha 1}$ at different variations of $\bar{q}_w$ has been derived from the calculation results in the form

$$\Delta K_{\alpha 1} = 26.6 (K_{q\tau})^{0.71} / (Re_f Pr_f^{0.6}) \tag{7.1}$$

for $Re_f = 10^4 - 10^6$, $Pr_f = 1-10$, $K_{q\tau} = 0-4000$, $x/d = 3.16-197$

$$\Delta K_{\alpha 1} = 1/[1 - 2.4 K_{q\tau}/Re_f Pr_f^{0.6}] - 1 \tag{7.2}$$

for $K_{q\tau} = -2000-0$, $Re_f = 10^4-10^5$, $Pr_f = 1-10$, $K_{q\tau}$ is found from (1.70). At $K_{q\tau} > 0 > \Delta K_{\alpha 1} > 0$ and at $K_{q\tau} < 0 < \Delta K_{\alpha 1} < 0$.

The empirical formulas for $\Delta K_{\alpha 2}$ and $\Delta K_{\alpha 3}$ at gas heating for different variations of T_w and G have the form:

(1) at increasing wall temperature

$$\Delta K_{\alpha 2} = \left(2 - 0.83 \frac{T_w}{T_f}\right)(10.4 - 19.2 \, \text{Re}_f \cdot 10^{-5}) \quad (7.3)$$
$$\times (K^*_{Tg} \cdot 10^4)^{1.836 - 0.664 \text{Re}_f \cdot 10^{-5}}$$

for $K^*_{Tg} = 0 - 0.4 \cdot 10^{-4}$, $\text{Re}_f = 7 \cdot 10^3 - 2.5 \cdot 10^4$, $T_w/T_f = 1-1.7$,

$$\Delta K_{\alpha 2} = \left(2 - 0.83 \frac{T_w}{T_f}\right)(4.6 - 1.46 \, \text{Re}_f \cdot 10^{-5}) \quad (7.4)$$
$$\times (K^*_{Tg} \cdot 10^4)^{1.605 - 0.1 \cdot \text{Re}_f \cdot 10^{-5}}$$

for $K^*_{Tg} = 0 - 0.4 \cdot 10^{-4}$, $\text{Re}_f = 2.5 \cdot 10^4 - 2 \cdot 10^5$, $T_w/T_f = 1-1.7$

$$\Delta K_{\alpha 2} = \frac{\exp\left[(1.45 - 7 \, \text{Re}_f \cdot 10^{-7})K^*_{Tg} \cdot 10^4\right] - 1}{\exp\left[0.4(2 - K^*_{Tg} \cdot 10^4)\text{Re}_f \cdot 10^{-5}\right]} \quad (7.5)$$

for $K^*_{Tg} = 0 - 1.1 \cdot 10^{-4}$, $\text{Re}_f = 8 \cdot 10^4 - 4.5 \cdot 10^5$, $T_w/T_f = 1-1.4$;

(2) at decreasing wall temperature

$$\Delta K_{\alpha 2} = -1.25 \left(2 - \frac{T_w}{T_f}\right)[1 - (0.325$$
$$+ 0.206 \, \text{Re}_f \cdot 10^{-5})]|K^*_{Tg} \cdot 10^4|^{0.105 \text{Re}_f 10^{-5}} - 0.27 \quad (7.6)$$

for $K^*_{Tg} = -0.4 \cdot 10^{-4} - -0.05 \cdot 10^{-4}$, $\text{Re}_f = 7 \cdot 10^3 - 2 \cdot 10^5$, $T_w/T_f = 1-1.7$,

$$\Delta K_{\alpha 2} = 1.25 \left(2 - \frac{T_w}{T_f}\right)(4.85 - 2.2 \, \text{Re}_f \cdot 10^{-5})K^*_{Tg} \cdot 10^4 \quad (7.7)$$

for $K^*_{Tg} = -0.05 \cdot 10^{-4} - 0$, $\text{Re}_f = 7 \cdot 10^3 - 2 \cdot 10^5$, $T_w/T_f = 1-1.6$,

$$\Delta K_{\alpha 2} = -\left(0.5 \frac{T_w}{T_f} - 0.42\right)[1 - \exp(4.6 \cdot 10^4 K^*_{Tg})] \quad (7.8)$$

for $K^*_{Tg} = -10^{-4} - 0$, $\text{Re}_f = 8 \cdot 10^4 - 5.2 \cdot 10^5$, $T_w/T_f = 1-1.6$.

In (7.3)–(7.8) the parameter is

$$K^*_{Tg} = \frac{\partial T_w}{\partial \tau} \frac{d}{T_w} \sqrt{\frac{\lambda}{c_p g G}} \quad (7.9)$$

(3) at increasing gas flow rate

$$\Delta K_{\alpha 3} = 0.004 \left(4.1 - 1.9 \frac{T_w}{T_f}\right) K_G^{(2.4 - 1.4 \text{Re}_f 10^{-5})} \quad (7.10)$$

for $K_G = 0-14$, $\text{Re}_f = 10^4 - 2.5 \cdot 10^5$, $T_w/T_f = 1-1.7$,

(4) at decreasing gas flow rate

$$\Delta K_{\alpha 3} = [(0.915 + 0.08 \mathrm{Re_f} \cdot 10^{-5})$$
$$\times |K_G|^{(0.25\mathrm{Re_f}10^{-5}-0.16)}]\left(0.66 + 0.275\frac{T_w}{T_f}\right) - 1 - A \qquad (7.11)$$

for $K_G = -30\text{--}-0.01$, $\mathrm{Re_f} = 10^4\text{--}2.5 \cdot 10^5$, $T_w/T_f = 1\text{--}1.7$, where

$$A = 0.045 \frac{T_w}{T_f} \bigg/ |K_G| \quad \text{at } \mathrm{Re_f} = (1.5 \ldots 6) \cdot 10^4$$

$$A = 0.08|K_G|^{0.5} \quad \text{at } \mathrm{Re_f} = (6 \ldots 12) \cdot 10^4$$

At $K_{Tg}^* > 0$ the value of $\Delta K_{\alpha 2} > 0$ and at $K_{Tg}^* < 0$ the value of $\Delta K_{\alpha 2} < 0$ (Fig. 7.2). At $K_G > 0$ the value of $\Delta K_{\alpha 3} > 0$ and at $K_G < 0$ the value of $\Delta K_{\alpha 3} < 0$ (Fig. 7.3), a $\Delta K_{\alpha 3}$ variation being the stronger, the less are $\mathrm{Re_f}$ and T_w/T_f.

These experiments have enabled one to evaluate the impact of a T_w variation on the turbulent flow structure. An assumption has been made that under unsteady conditions a steady turbulent thermal diffusion distribution is preserved, and an empirical multiplier B has been included into the term for the dimensionless distances from the wall η; at $K_{Tg}^* > 0$ $B > 0$. The values of λ_T/λ (λ_T is the turbulent thermal conductivity) increase with T_w

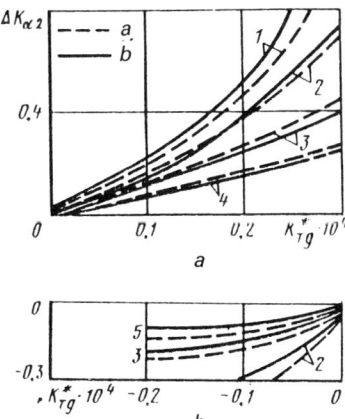

Figure 7.2 Comparison of $\Delta K_{\alpha 2}$ vs K_{Tg}^* for sharp (a) and smooth (b) variations of heat load at gas heating: 1) $\mathrm{Re_f} = (1\text{--}2) \cdot 10^4$, $T_w/T_f = 1.1\text{--}1.3$; 2) $\mathrm{Re_f} = (6\text{--}7) \cdot 10^4$, $T_w/T_f = 1.1\text{--}1.3$; 3) $\mathrm{Re_f} = (7\text{--}9) \cdot 10^4$, $T_w/T_f = 1.3\text{--}1.5$; 4) $\mathrm{Re_f} = (1.2\text{--}1.5) \cdot 10^5$, $T_w/T_f = 1.5\text{--}1.7$; 5) $\mathrm{Re_f} = (7\text{--}9) \cdot 10^4$, $T_w/T_f = 1.5\text{--}1.7$.

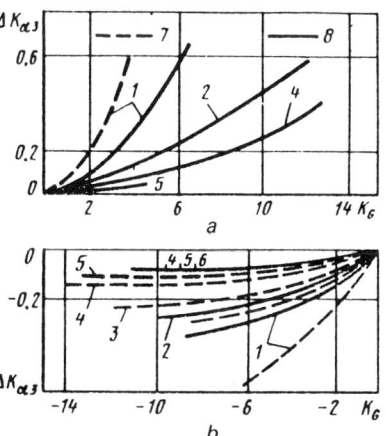

Figure 7.3 $\Delta K_{\alpha 3}$ vs K_G for different $\mathrm{Re_f}$ and T_w/T_f at increasing (a) and decreasing (b) flow rates of a heated gas: 1–6) $\mathrm{Re_f} \cdot 10^{-4} = 1\text{--}2, 2\text{--}3, 3\text{--}4, 4\text{--}5, 5\text{--}6, 6\text{--}7$; 7) $T_w/T_f = 1.2\text{--}1.3$; 8) $T_w/T_f = 1.4\text{--}1.5$.

3–4 times in the wall region at a moderate K_α growth and in the flow core, by 20–50% as much (Fig. 7.4).

The relations obtained for $\Delta K_{\alpha 1}$ and $\Delta K_{\alpha 2}$ offer finding a relationship between them during time variations of T_w and q_w at gas heating in a tube. As seen from (7.1), (7.3) and (7.4), with increasing heat load at $\text{Re}_f = \text{const}$

$$\Delta K_{\alpha 1} = C_1 (K_{q\tau})^{0.71} \qquad (7.12)$$

$$\Delta K_{\alpha 2} = C_2 (K_{Tg}^*)^n \qquad (7.13)$$

where $n = 1.796$–1.405 at Re_f ranging from $7 \cdot 10^3$ to $2 \cdot 10^5$.

Considering the relations for $K_{q\tau}$ and K_{Tg}^* as well as the expression for $G = 0.25 \, d \, \mu_f \, \text{Re}_f$ yields that under other conditions being equal

$$\Delta K_{\alpha 1} / \Delta K_{\alpha 2} = C d^m \qquad (7.14)$$

where $m = 0.522$–0.8075 for $\text{Re}_f = 7 \cdot 10^3$–$2 \cdot 10^5$. Thus, $\Delta K_{\alpha 1} / \Delta K_{\alpha 2}$ is the higher, the greater is the tube diameter. For 10–20 mm dia tubes used in heat exchange apparatuses $\Delta K_{\alpha 1} \ll \Delta K_{\alpha 2}$ (Fig. 1.3), and it may be assumed that $\Delta K_{\alpha 1} = 0$, i.e. a difference of K_α from 1 is specified by altering a turbulent flow structure. For large d (e.g. in calculating unsteady heat transfer in gas pipelines) $\Delta K_{\alpha 1}$ must be taken into account.

At a constant hot gas flow rate and heating of walls the empirical formula for $\Delta K_{\alpha 2}$ at different wall temperature variations is of the form:

$$\Delta K_{\alpha 2} = \{[14.97(T_w/T_f)^3 - 16.07(T_w/T_f)^2$$

$$-0.526(T_w/T_f) + 3.193](\text{Re}_f \cdot 10^{-5})^{1.15 - 3T_w/T_f}$$

$$+ 46.77(T_w/T_f)^3 - 119.1(T_w/T_f)^2 + 99.09(T_w/T_f)$$

$$-27.08\} \cdot K_{Tg}^* \cdot 10^5 \qquad (7.15)$$

for $K_{Tg}^* = 0$–$2 \cdot 10^{-4}$, $\text{Re}_f = 3.2 \cdot 10^4$–$2 \cdot 10^5$, $T_w/T_f = 0.6$–1.

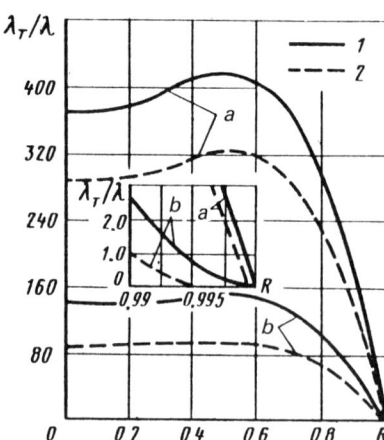

Figure 7.4 Radius distribution of turbulent heat conduction: 1) calculation with regard to unsteadiness; 2) calculation to a quasi-stationary approximation; a) $\text{Re}_f = 2.5 \cdot 10^5$, $T_w/T_f = 1.12$; $K_{T_g} = (\partial T_w/\partial \tau)(d/(T_w - T_f)_0) \sqrt{\lambda / c_p g G} = 1.26 \cdot 10^{-4}$, $K_\alpha = 1.29$; b) $\text{Re}_f = 5.5 \cdot 10^4$, $T_w/T_f = 1.16$, $1.1 \cdot 10^{-4}$, $K_\alpha = 1.45$ $(T_w - T_f)_0$ is the finite temperature head.

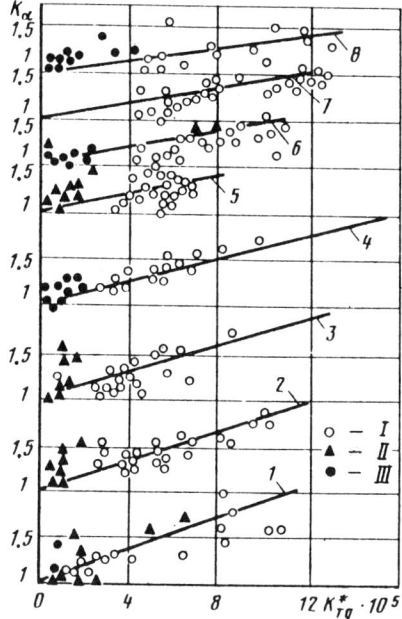 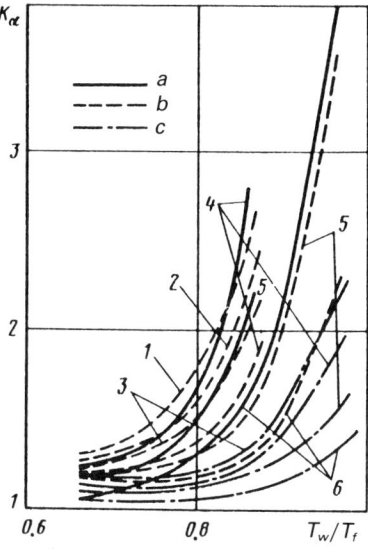

Figure 7.5 K_α vs K^*_{Tg} at gas cooling ($T_w/T_f = 0.7$–0.8) and at different Re_f: *1–8*) $Re_f \cdot 10^{-4} = 3.2$–4, 4–5, 5–6.25, 6.25–7.83, 7.83–9.83, 9.83–12.25, 12.25–15.5, 15.5–20. *I, II, III*) tubes with $d = 8.65$, 42.8, 9.82 mm.

Figure 7.6 K_α vs T_w/T_f at different Re_f and K^*_{Tg} at gas cooling: *a*) $Re_f = (3.2$–$4) \cdot 10^4$; *b*) $Re_f = (6.25$–$7.83) \cdot 10^4$; *c*) $Re_f = (1.55$–$2) \cdot 10^5$; for *1–6*, $K^*_{Tg} \cdot 10^5 = 6, 5, 4, 3, 2, 1$,

At wall heating ($K^*_{Tg} > 0$) $\Delta K_{\alpha 3} > 0$. The values of $\Delta K_{\alpha 2}$ are the higher, the greater is K^*_{Tg} (Fig. 7.5). The impact of unsteadiness grows with decreasing Re_f (Fig. 7.5) and T_w/T_f (Fig. 7.6). The results on the experiments with periodic hot gas flow rate despite a difference in the initial conditions of each cycle, agree with formula (7.15) and support the possibility to employ expression (7.15) for calculation of heat transfer at arbitrary T_w variations.

The empirical formulas for liquid heating at different T_w and G variations are of the following form.

(1) At varying heat load the quantity $|\Delta K_{\alpha 2}|$ is the larger, the greater is $|K^*_{Tg}|$ and the less is Re_f, does not depend on Pr_f and is generalized at $Pr_f = 3$–10 by the formulas:

$$\Delta K_{\alpha 2} = (1.72 \cdot 10^6/Re_f^{0.303})K^*_{Tg} \tag{7.16}$$

for $Re_f = 4 \cdot 10^3$–$2 \cdot 10^4$, $K^*_{Tg} = 0$–$0.7 \cdot 10^{-5}$,

$$\Delta K_{\alpha 2} = (8.29 \cdot 10^9/Re_f^{1.16})K^*_{Tg} \tag{7.17}$$

for $Re_f = 2 \cdot 10^4 – 10^5$, $K^*_{Tg} = 0 – 0.7 \cdot 10^{-5}$,

$$\Delta K_{\alpha 2} = (1 - 1.72 \cdot 10^6 \, K^*_{Tg}/Re_f^{1.16})^{-1} - 1 \qquad (7.18)$$

for $Re_f = 4 \cdot 10^3 – 2 \cdot 10^4$, $K^*_{Tg} = -0.3 – 10^{-5} – 0$,

$$\Delta K_{\alpha 2} = (1 - 8.29 \cdot 10^9 \, K^*_{Tg}/Re_f^{1.16})^{-1} - 1 \qquad (7.19)$$

for $Re_f = 2 \cdot 10^4 – 5 \cdot 10^4$, $K^*_{Tg} = -0.3 \cdot 10^5 – 0$.

For liquids the values of $\Delta K_{\alpha 1}$ and $\Delta K_{\alpha 2}$ are comparable. The ratio $\Delta K_{\alpha 2}/\Delta K_{\alpha 1}$ decreases with increasing a tube diameter at the same $\partial T_w/\partial \tau$, T_w/T_f, Re_f and Pr_f, therefore, K_α as a function of $K_{q\tau}$ or K^*_{Tg} is not unique for different-diameter tubes, and two parameters of thermal unsteadiness must be included.

At the same values of K^*_{Tg} determined by formula (1.83) and of Re_f, the values of $\Delta K_{\alpha 2}$ for a liquid and gas (at T_w/T_f close to 1) virtually coincide (Fig. 7.7) although the volumetric expansion coefficient ratio may amount to 40. This supports the validity of the model (Section 1.3) for the impact of wall temperature variations on the turbulent flow structure and unsteady heat transfer which are the most pronounced, the greater is $\partial T_w/\partial \tau$ and β_w.

The conducted experiments and their analysis have evidenced that the effect of altering a turbulent flow structure on unsteady heat transfer is substantial both for gases and for liquids.

(2). The values of $\Delta K_{\alpha 3}$ obtained from the experimental values of K_α and of $\Delta K_{\alpha 1}$ and $\Delta K_{\alpha 2}$ found at $G = $ const and at $Pr_f = 3–12$, $Pr_f/Pr_w = 1–4$, $x/d = 6–160$ are generalized by the formulas (Fig. 7.8):

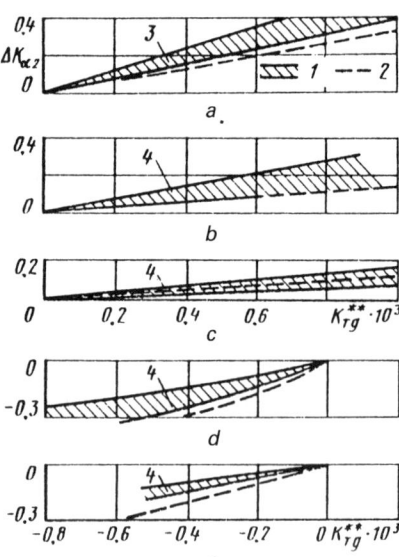

Figure 7.7 Comparison of the data for the effect of altering a turbulent flow structure on unsteady heat transfer at gas and liquid heatings: *1*) water; $Pr_f = 4–12$ (*3*), $Pr_f = 3–12$ (*4*); *2*) air; $Pr_f = 0.72$; $K^*_{Tg} = (\partial T_w/\partial \tau) \beta_w \sqrt{d/g}$; for *a, b, c, d, e*, $Re_f = 8 \cdot 10^3, 2.5 \cdot 10^4, 5 \cdot 10^4, 8 \cdot 10^3, 2.4 \cdot 10^4$.

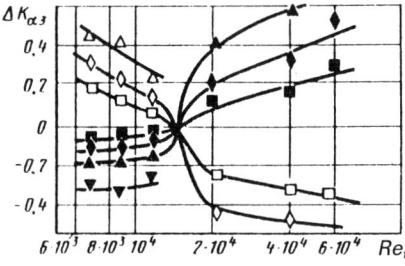

Figure 7.8 $\Delta K_{\alpha 3}$ vs Re_f and K_G at increasing ($K_G > 0$) and decreasing ($K_G < 0$) flow rate of a heated liquid: $\triangledown$, $\blacktriangle$, $\blacklozenge$, $\blacksquare$, $\square$, $\diamond$, $\triangle$, K_G, = 400, 200, 100, 50, -50, -100, -200.

for $Re_f = (6\text{–}12) \cdot 10^3$ $K_G = 0\text{–}400$

$$\Delta K_{\alpha 3} = (6 \cdot 10^{-9} K_G + 5.6 \cdot 10^{-6}) Re_f - 7 \cdot 10^{-4} K_G$$
$$- 0.071 \qquad (7.20)$$

for $Re_f = (6\text{–}12) \cdot 10^3$ $K_G = -200\text{–}0$

$$\Delta K_{\alpha 3} = (9.3 \cdot 10^{-8} K_G - 2 \cdot 10^{-5}) Re_f - 2.4 \cdot 10^{-2} K_G$$
$$+ 0.236 \qquad (7.21)$$

for $Re_f = (12\text{–}20) \cdot 10^3$ $K_G = -100\text{–}200$,

$$\Delta K_{\alpha 3} = (2.43 \cdot 10^{-2} K_G - 5.67 \cdot 10^{-2}) Re_f^{0.22}$$
$$- (3.57 \cdot 10^{-2} K_G - 0.83) \qquad (7.22)$$

for $Re_f = (20\text{–}60) \cdot 10^3$ $K_G = 0\text{–}200$

$$\Delta K_{\alpha 3} = (3.91 \cdot 10^{-8} K_G + 2.173 \cdot 10^{-6}) Re_f$$
$$+ 1.13 \cdot 10^{-3} K_G^{-0.116} \qquad (7.23)$$

for $Re_f = (20\text{–}60) \cdot 10^3$ $K_G = -100\text{–}0$

$$\Delta K_{\alpha 3} = (-5 \cdot 10^{-9} K_G - 2.75 \cdot 10^{-6}) Re_f$$
$$+ 2.8 \cdot 10^{-3} K_G - 0.07 \qquad (7.24)$$

At $Re_f = (1.5\text{–}6) \cdot 10^4$ $\Delta K_{\alpha 3} > 0$ at $K_G > 0$ and $\Delta K_{\alpha 3} < 0$ at $K_G < 0$. At decreasing Re_f, the effect of high-speed unsteadiness on heat transfer reduces and then becomes opposite: at flow acceleration, heat transfer decreases while at flow deceleration it increases, as compared to the quasi-stationary one. Since in calculations the values of T_w and $\partial T_w/\partial \tau$ (as well as of q_w and $\partial q_w/\partial \tau$) are beforehand unknown, the problem is being solved by the successive approximation method. To a first approximation, heat transfer coefficients are found using the quasi-stationary relations. Then the values of T_w, $\partial T_w/\partial \tau$, K_{Tg}^*, q_w, $\partial q_w/\partial \tau$, K_{qT} and of the unsteady heat transfer coefficient are estimated to a first approximation. As a result, the following approximations may be made to solve this problem.

It should be noted that the empirical design formulas given in the present section offer, with prescribed calculating accuracy, determination of permissible rates of varying the parameters ($\partial T_w/\partial \tau$, $\partial q_w/\partial \tau$, $dG/d\tau$), and the fields of applicability of the quasi-stationary relations for the transfer coefficient. For instance, Figure 7.9 plots the limiting values of the parameters K^*_{Tg} vs Re_f and T_w/T_f, at which the values of $\Delta K_{\alpha 2}$ do not exceed the assigned ones at hot gas cooling in tubes.

The conducted experiments have witnessed that under turbulent flow operating conditions, a difference of the unsteady heat transfer coefficient from the quasi-stationary one is specified not by the laws for boundary conditions but only by the rates of their change, i.e. by the first derivatives of flow rate, wall temperature or wall heat flux density. The appropriate dimensionless parameters K_{qT}, K^*_{Tg} and K_G responsible for heat transfer coefficient change under unsteady conditions are obtained. These experiments and their analysis have evidenced that the effect of altering a turbulent flow structure on unsteady heat transfer is substantial both for gases and for liquids. Therefore, unsteady heat transfer calculations using a concept of a quasi-stationary turbulent structure yield inadmissible errors.

Thus, the experimentally checked generalizing relations are obtained to calculate the coefficient of unsteady heat transfer involving gas and liquid tube flows under different unsteady impacts encountered in practice over a wide range of the parameters. In particular, these relations offer, with assigned calculating accuracy, specifying the limits of applicability of the quasi-stationary methods of calculating unsteady thermal processes.

7.3 UNSTEADY HEAT TRANSFER IN HELICAL TUBE BUNDLES

Unsteady heat transfer in a helical tube bundle was investigated at constant mass flow rate and at varying heat release in the tube walls of a bundle. Heat release in a bundle was varied abruptly. In the case of increasing heat release it was absent at the initial moment of the process ($q_{v1} = 0$); in the case of decreasing heat release the electric load was fully removed at the initial moment of the process ($q_{v2} = 0$). Several operating conditions with a smooth change of heat release in a bundle were examined.

In experiments, the main parameters were varied over the following ranges: $Re_f = 5 \cdot 10\text{--}5 \cdot 10^4$, maximum wall temperature $T_w = 670\text{--}820$ K, temperature factor $T_w/T_f = 1\text{--}1.4$, maximum rate of varying a wall temperature $|\partial T_w/\partial \tau| = 50$ K/s, air pressure $(2.9\text{--}10.7) \cdot 10^7$ Pa.

Experiments were made in the following fashion. A prescribed air flow rate via a tube bundle was set. Then the electric load was supplied, a bundle was heated. As far as a bundle was heated, a rate of increasing a bundle

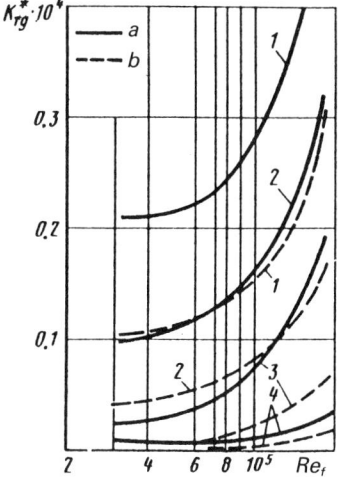

Figure 7.9 Limiting values of K^*_{fg}, at which ΔK_α does not exceed 10% (*a*) and 5% (*b*); *1–4*) T_w/T_f = 0.65, 0.75, 0.85, 0.95.

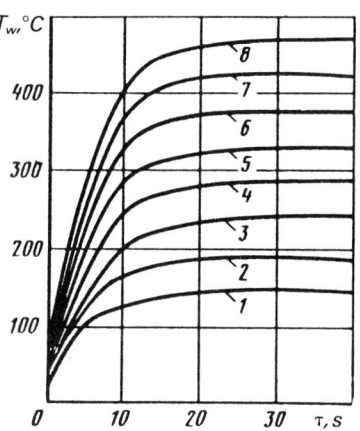

Figure 7.10 Time variation of a wall temperature of a tube bundle at sharply increasing heat release for $\mathrm{Re}_f = (2.7-3.5)\cdot 10^4$, $T_w/T_f = 1-1.28$: *1–8*) x/d_{eq} = 7.12, 13.1, 24.9, 48.6, 78.4, 108, 137, 167.

temperature decreased (Fig. 7.10), and a heat transfer process was stabilized. The stabilization time of the wall temperature amounted to 15–20 s.

In increasing a wall temperature at the initial time moments, K_α attained 2–3, and then as far as a tube bundle was heated, heat transfer was stabilized and K_α approached 1 (Fig. 7.11). The stabilization of K_α was advanced the quicker, the greater was Re_f. The stabilization time of the heat transfer coefficient (which was characterized by $K_\alpha \leq 105$) was 8–12 s. After the experimental set-up had achieved a steady heating regime, the electric load was removed. The wall temperature fell with a decreasing rate the stronger, the greater was x/d_{eq}. It asymptotically approached a cooling air temperature. The wall temperature was stabilized earlier at the bundle inlet than at its outlet. The stabilization time of the wall temperature decreased with increasing Re_f. It was somewhat larger than the one of a wall temperature at increasing heat load and amounted to 20–25 s.

At the initial moment of an unsteady process, when the electric load was removed, $K_\alpha = 0.5-0.7$, and then as far as a tube bundle was cooled off heat transfer was stabilized, and K_α approached 1. It should be noted that as far as a tube bundle was cooled off the heat flux density (proportional to a rate of falling a wall temperature) and the temperature head between a wall and flow decreased. This inevitably augmented an error in determining the heat transfer coefficient.

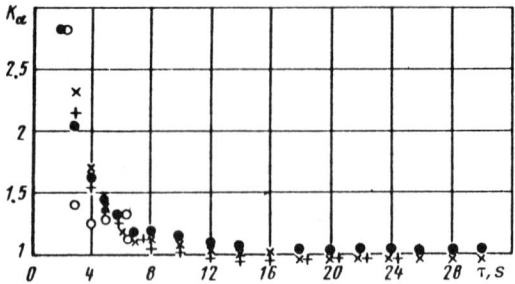

Figure 7.11 Time variation of the coefficient K_α at sharply increasing heat release in bundle tube walls: ○, ×, +, ◁, $x/d_{eq} = 7.12$, 13.1, 24.9, 48.6.

As emphasized in Chapter 6, the experiment was aimed at finding K_α as a function of $\partial T_w/\partial \tau$ or thermal unsteadiness parameter K_{Tg}^* (determined by formula (1.82)) as a function of the parameters Re_f, T_w/T_f, x/d_{eq}, i.e.

$$K_\alpha = 1 + \Delta K_{\alpha 2} = \frac{Nu_f\left(Re_f \frac{T_f}{T_f}, \frac{x}{d_{eq}}, K_{Tg}^*\right)}{Nu_{f.qs}\left(Re_f, \frac{T_w}{T_f}, \frac{x}{d_{eq}}\right)} = f\left(K_{Tg}^*, Re_f, \frac{T_w}{T_f}, \frac{x}{d_{eq}}\right) \quad (7.25)$$

The values of $\alpha(x, \tau)$ were determined for eight cross-sections of a bundle at the distances $x/d_{eq} = 7.12$, 13.1, 24.9, 48.6, 78.9, 108, 137, 167 from the initial moment of heating. The quasi-stationary values of the mean heat transfer coefficient were found for the considered bundles by the relation [39]:

$$Nu_f = 0.035 \, Re_f^{0.75}\left(1 + \frac{\pi^2}{0.5 \dfrac{S}{d}\dfrac{S}{d_{eq\infty}}}\right)$$

$$\left[1 + \frac{1.3}{\left(\dfrac{S}{d}\dfrac{S}{d_\infty}\right)^{0.6}}\right]\left(\frac{T_w}{T_f}\right)^{-n} \quad (7.26)$$

where

at $S/d \leq 4.15$ $\quad\quad n = 0$ $\quad\quad\quad\quad\quad\quad\quad\quad\quad\quad\quad\quad (7.27)$

at $4.15 < S/d < 12.45$ $\quad n = 0.0663 \dfrac{S}{d} - 0.275$ $\quad\quad (7.28)$

at $S/d \geq 12.45$ $\quad\quad n = 0.55$ $\quad\quad\quad\quad\quad\quad\quad\quad\quad (7.29)$

Nu_f and Re_f were determined in terms of the flow temperature and the equivalent bundle diameter d_{eq}. This bundle was composed of an infinite number of tubes. Formula (7.26) was derived from the steady heat transfer experi-

ments made on the same helical tube bundles and was valid for $\text{Re}_f = 6 \cdot 10^3 - 10^5$ and $T_w/T_f = 1-1.4$. Quasi-stationary values of the mean heat transfer coefficient for the studied helical tube bundles may be also estimated by the following formulas:
for $S/d = 12.45$

$$\text{Nu}_f = 0.0232 \, \text{Re}_f^{0.8} (T_w/T_f)^{-0.55} \qquad (7.30)$$

for $S/d = 4.15$

$$\text{Nu}_f = 0.0851 \, \text{Re}_f^{0.7} \qquad (7.31)$$

Since in experiments, K_{Tg}^*, Re_f and T_w/T_f were interdependent, that relation (7.25) be derived, the experimental points were classified according to the following ranges of the parameters: $\text{Re}_f \cdot 10^{-4} = 1-1.5$, 1.5–1.8, 1.8–2, 2–2.3, 2.3–2.7, 2.7–3, 3–3.5, 3.5–4, 4–4.5, 4.5–5 and $T_w/T_f = 1-1.1$, 1.1–1.2, 1.2–1.3, 1.3–1.4. Then for each range the dependences of K_α on K_{Tg}^* were plotted, and the averaging relations were referred to the mean values of Re_f and T_w/T_f in their considered ranges.

Figures 7.12 and 7.13 plot K_α vs K_{Tg}^* (7.25) at different x/d_{eq} and Re_f for two helical tube bundles over one range of $T_w/T_f = 1.1-1.2$. Despite the fact that the number of the experimental points is not so large, from this

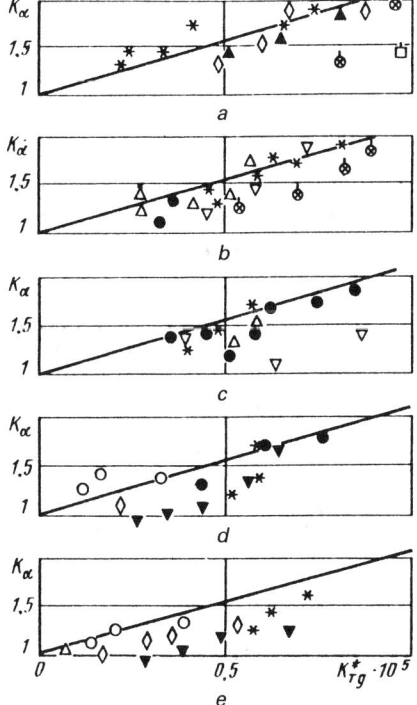

Figure 7.12 K_α vs K_{Tg}^* with increasing heat load at different Re_f, x/d_{eq} and $T_w/T_f = 1.1-1.2$ for helical tube bundles: ○, △, ◇, ●, *, ⊗, □, ▽, ▲, $\text{Re}_f \cdot 10^{-4} = 1-1.5$, 1.5–1.8, 1.8–2, 2–2.3, 2.3–2.7, 2.7–3, 3–3.5, 3.5–4, 4–4.5, 4.5–5 for a bundle with $S/d = 4.15$, respectively; *, ⊗, ⊡, $\text{Re}_f \cdot 10^{-4} = 2.3-2.7$, 2.7–3.3, 3.3–3.5 for a bundle with $S/d = 12.45$; a, b, c, d, e) $x/d_{eq} = 4.96$, 78.8, 108, 132, 167, respectively.

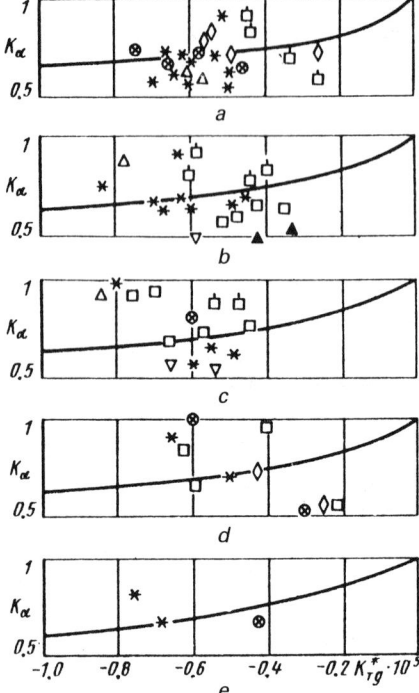

Figure 7.13 K_α vs K_{Tg}^* with decreasing heat load for helical tube bundles at $T_w/T_f = 1.1$–1.2. The notations are the same as in Fig. 7.12.

figure it is seen that a deviation of K_α from 1 augments with $|K_{Tg}^*|$. Within a scatter of the points the effect of the tube twisting pitch S/d on K_α is not observed. Therefore, the experimental points referring to two versions of bundles as well as to different x/d_{eq} and Re_f have been analyzed simultaneously. As seen from Figures 7.12 and 7.13, in the case of the data processed by formula (7.25), the contribution of the distance x/d_{eq} from the bundle inlet to K_α does not manifest itself both at increasing and at decreasing heat load. Similar relations were derived for other ranges of T_w/T_f.

Figures 7.14 and 7.15 plot all the points obtained at increasing and decreasing heat load with K_α as a function of K_{Tg}^* for different T_w/T_f and Re_f. The appropriate averaging dependences of K_α on K_{Tg}^* for different T_w/T_f are shown in Fig. 7.16. A difference of K_α from 1 at increasing heat load ($K_\alpha > 1$) is the greater, the less is T_w/T_f while at decreasing heat load ($K_\alpha < 1$) the temperature factor does not affect K_α as a function of K_{Tg}^*.

Inspection of Figs. 7.14 and 7.15 shows that in the case of the data processed by formula (7.25) the effect of Re_f on unsteady heat transfer is not revealed over the analyzed range of the Re_f number. This may be attributed to the fact that the parameter K_{Tg}^* incorporates a mass flow velocity, i.e. the parameter K_{Tg}^* decreases with increasing Re_f at the same values of

$\partial T_w/\partial \tau$. This indirectly takes into account the reducing effect of thermal unsteadiness on heat transfer due to increasing Reynolds number.

$$K_\alpha = 1 + \Delta K_{\alpha 2} = 1 + (K^*_{Tg} \, 10^5) \frac{2.4}{(T_w/T_f)^{6.2}} \tag{7.32}$$

The experimental data are generalized by the relation to elevate heat load at $Re_f = (1-5) \cdot 10^4$, $T_w/T_f = 1-1.4$ and $K^*_{Tg} = 0-(3.6-2.3 \, T_w/T_f) \cdot 10^{-5}$. As seen from Fig. 7.16, at the same values of the characteristic parameters K^*_{Tg}, Re_f and T_w/T_f in helical tube bundles a difference of the unsteady heat transfer coefficient from the quasi-stationary one is greater than in a round tube.

With decreasing heat load, at $Re_f = (1-5) \cdot 10^4$, $T_w/T_f = 1-1.4$ and $K^*_{Tg} = -1.2 \cdot 10^{-5}$, the experimental data are generalized by the relation

$$K_\alpha = 1 + \Delta K_{\alpha 2} = 1 - 0.412(1 - e^{1.897 \, K^*_{Tg} \cdot 10^5}) \tag{7.33}$$

As seen from Fig. 7.15 and formula (7.33), with lowering heat load over the investigated range of the parameters, the temperature factor does not exert an influence on K_α. As in the case of increasing heat load, a deviation of K_α from 1 is greater than in a round tube.

These investigations have illustrated that at varying a helical tube wall temperature a deviation of the unsteady heat transfer coefficient from the quasi-stationary one may be pronounced. It is the more substantial, the greater is a rate of varying a wall temperature $\partial T_w/\partial \tau$. Generalizing relations (7.32)

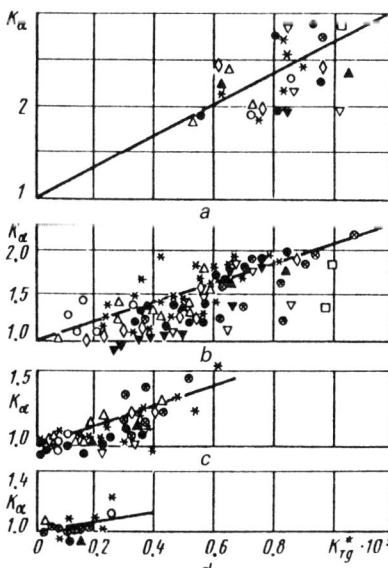

Figure 7.14 K_α vs K^*_{Tg} at increasing heat load and at different Re_f and T_w/T_f for helical tube bundles with $S/d = 4.15$ and 12.45: a, b, c, d) $T_w/T_f = 1-1.1$, $1.1-1.2$, $1.2-1.3$, $1.3-1.4$. The remainder notations are the same as in Fig. 7.12.

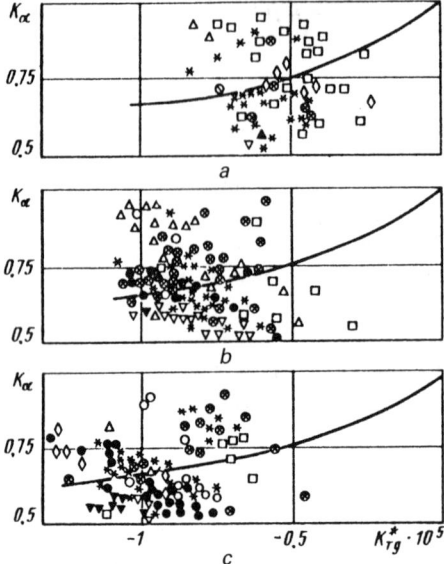

Figure 7.15 K_α vs K^*_{Tg} at decreasing heat load and at different Re_f and T_w/T_f for helical tube bundles with $S/d = 4.15$ and 12.45: a, b, c) $T_w/T_f = 1.1-1.2$, $1.2-1.3$, $1.3-1.4$. The remainder notations are the same as in Fig. 7.12.

and (7.33) provide calculation of the unsteady heat transfer coefficient in a helical tube bundle if the variations of boundary conditions or the time variations of a wall temperature are known. These formulas enable one to specify the boundaries of the usability of quasi-stationary relations at a prescribed accuracy of calculating the unsteady heat transfer coefficient. At a permissible error associated with defining a difference of the unsteady heat transfer

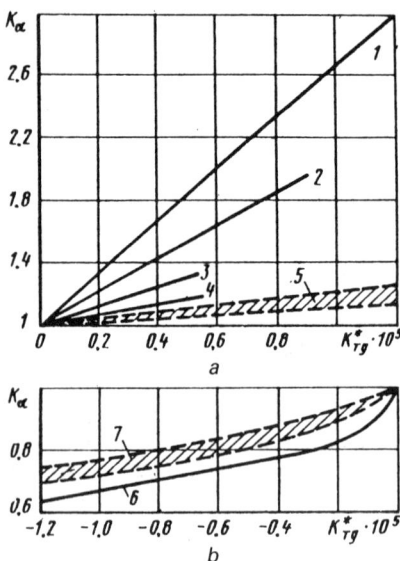

Figure 7.16 Comparison of K_α as a function of K^*_{Tg} for helical tube bundles ($S/d = 4.15$ and 12.45) and for a tube: a) increasing heat load; 1–4) helical tube bundles for $T_w/T_f = 1-1.1$, $1.1-1.2$, $1.2-1.3$, $1.3-1.4$, respectively, at $Re_f = (1-5) \cdot 10^4$; 5) tube, $Re_f = (1-5) \cdot 10^4$, $T_w/T_f = 1.1-1.5$; b) decreasing heat load; 6) helical tube bundles at $Re_f = (1-5) \cdot 10^4$, $T_w/T_f = 1-4.4$; 7) tube, $Re_f = (1-5) \cdot 10^4$, $T_w/T_f = 1.1-1.5$.

coefficient from the quasi-stationary one ΔK_α, the limiting values of the thermal unsteadiness parameter must not exceed those of the quantity

$$K^*_{Tg} = \Delta K_\alpha \left(\frac{T_w}{T_f}\right)^{6.2}/2.4 \qquad (7.34)$$

at increasing heat load

$$K^*_{Tg} = 5.28 \cdot 10^{-6} \ln(1 - 2.42/\Delta K_\alpha) \qquad (7.35)$$

at decreasing heat load.

It should be noted that the thermal unsteadiness parameter K^*_{Tg}, (1.82), which was for the first time proposed in [26] and used, in the present book, for correlating the experimental data, is also successfully adopted to generalize the data on unsteady heat transfer in other complex-geometry channels. In [23, 31], this parameter was employed to generalize the data on heat transfer between a gas and a profile of the nozzle blades of the axial turbine.

Figures 7.17 and 7.18 gives a comparison of the data of the present section and the available experimental results on unsteady heat transfer in channels of different geometry. G. A. Dreitser and V. V. Balashov's data on unsteady heat transfer in a flat channel at one-sided heating are mentioned. A channel 40 mm wide, 2.5 mm high and 230 mm long had no preliminary hydrodynamic stabilization length. One wide surface of the channel was heated. The experimental data for increasing heat load are generalized by the relation

$$K_\alpha = 1 + \Delta K_{\alpha 2} = 1 + \left(\frac{460}{Re_f^{0.51}} + \frac{x}{d_{eq}}\frac{8.13}{Re_f^{0.46}}\right) K^*_{Tg} \, 10^4 \qquad (7.36)$$

valid at $Re_f = (1-8.8) \cdot 10^4$, $K^*_{Tg} = (0-0.22) \cdot 10^{-4}$, $T_w/T_f = 1-1.7$ and $x/d_{eq} = 10.4-44$. At lowering heat load the generalizing formula is obtained

$$K_\alpha = 1 + \Delta K_{\alpha 2} = \exp(0.813 \, K^*_{Tg} \, 10^4) \qquad (7.37)$$

which is valid at $Re_f = (1-8.5) \cdot 10^4$, $K^*_{Tg} = -0.22 \cdot 10^{-4}-0$, $T_w/T_f = 1-1.7$ and $x/d_{eq} = 10.4-44$.

These figures also display the data on unsteady heat transfer between a hot gas and a nozzle blade profile [31]. In experiments, a hot gas temperature in front of the blades abruptly increased and the blades had undergone heating. The temperature unsteadiness was initiated almost abruptly at a constant gas flow rate in front of a grid. In experiments, a 3-6 fold increase of the unsteady heat transfer coefficient was attained, as compared to the quasi-stationary one at the beginning of a process. The generalizing relation was derived

$$K_\alpha = 1 + \{\exp[a - b(K^*_{Tg} \cdot 10^5)^{-c}]\} m Re_f^n \qquad (7.38)$$

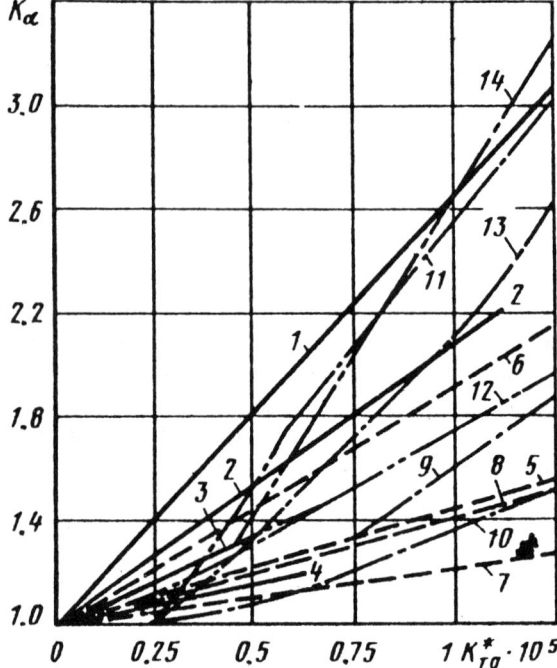

Figure 7.17 Comparison of K_α as a function of K_{Tg}^* for different channels at increasing wall temperature: *1–4*) in-line helical tube bundles at $Re_f \cdot 10^{-4} = 1–5$ for $T_w/T_f = 1.05$, 1.15, 1.25, 1.35, respectively; *5–8*) flat channel with one-sided heating with no hydrodynamic stabilization length for $T_w/T_f = 1–1.7$; *5–6*) $x/d_{eq} = 10.4$ and 44 for $Re_f = 10^4$; *7, 8*) the same for $Re_f = 5 \cdot 10^4$; *9, 10*) nozzle blade back at 0.3 chord distant from the outlet edge for $Re_f = 3 \cdot 10^5$ and $1.4 \cdot 10^6$ [31]; *11*) remainder part of the blade back for $Re_f = 3 \cdot 10^5$–$1.4 \cdot 10^6$ [31]; *12*) nozzle blade tray at 0.3 chord distant from the outlet edge for $Re_f = 3 \cdot 10^5$–$1.4 \cdot 10^6$ [31]; *13*) remainder part of the blade tray for $Re_f = 3 \cdot 10^5$ and $1.4 \cdot 10^6$ [31].

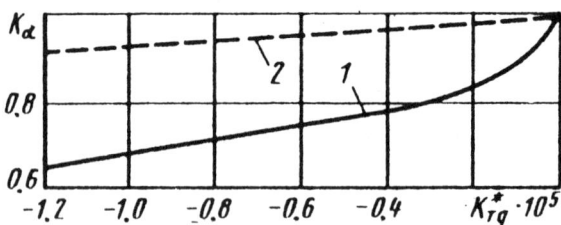

Figure 7.18 Comparison of K_α as a function of K_{Tg}^* for different channels with decreasing wall temperature: *1*) in-line helical tube bundle at $Re_f \cdot 10^{-4} = 1–5$, $T_w/T_f = 1–1.4$; *2*) flat channel with one-sided heating in the absence of the hydrodynamic stabilization length for $x/d_{eq} = 10.4–44$, $Re_f \cdot 10^{-4} = 1.25–8.5$, $T_w/T_f = 1–1.7$.

which was valid for $K^*_{\mathrm{T}g} = 0\text{–}3.6 \cdot 10^{-5}$, $\mathrm{Re}_f = 3 \cdot 10^5\text{–}1.4 \cdot 10^6$. A velocity in the minimum cross-section and a gas stagnation temperature were taken as characteristic sizes. A profile chord length served as a characteristic dimension. The following values of the constants were obtained: for a wall at 0.3 chord distant from the outlet edge $a = 2.52$, $b = 3.29$, $c = 0.55$, $m = 78.87$, $n = -0.327$, for the remainder parts of a wall 1.91, 1.46, 0.81, 1.00, 0, respectively; for a tray at 0.3 chord distant from the outlet edge: 1.51, 1.8, 0.59, 1.00, 0, respectively; for the remainder parts of a tray: 1.99, 1.6, 0.88, 0.048, 0.224.

As seen from Figs. 7.17 and 7.18, the experimental data for different channels both at increasing and at decreasing heat load greatly diverge although in all the cases a deviation of K_α from 1 grows with $|K^*_{\mathrm{T}g}|$. The effect of such parameters as x/d_{eq}, Re_f and T_w/T_f on the relation $K_\alpha = f(K^*_{\mathrm{T}g})$ is different. For example, this relation is affected by the temperature factor T_w/T_f and not affected by Re_f and x/d_{eq} in a helical tube bundle at increasing heat load. In a flat channel, vice versa, T_w/T_f does not influence K_α as a function of $K^*_{\mathrm{T}g}$ but it is affected by x/d_{eq} and Re_f. The influence of the Reynolds number is different in different channels. All this emphasizes that the processes under study are complex, and the experimental data on the channels of one geometry cannot be reliably extended to those of another.

CHAPTER
EIGHT

METHODS OF CALCULATING UNSTEADY HEAT AND MASS TRANSFER. SOME RECOMMENDATIONS

8.1 CALCULATION OF UNSTEADY HEAT TRANSFER AT UNIFORM HEAT SUPPLY TO A BUNDLE

The experimental investigations carried out offer determining the coefficient of heat transfer from helical tube bundles in the longitudinal flow under unsteady conditions as well as allowing for it as a function of a rate of varying boundary conditions. The earlier experiments in round tubes make it possible also to determine unsteady heat transfer coefficients inside helical tubes with sufficient accuracy. Since heat transfer inside helical tubes slightly differs from the one inside round tubes, the data on K_α for round tubes may be used to calculate unsteady heat transfer inside helical tubes.

Under unsteady conditions the heat transfer coefficient has the same meaning as under steady conditions, however, it depends not only on the numbers responsible for steady heat transfer (Re_f, Pr_f, x/d, T_w/T_f for gases and ρ_w/ρ_f, μ_w/μ_f, c_{pw}/c_{pf}, λ_w/λ_f for liquids) but also on the numbers specified by a rate of varying boundary conditions, namely, wall temperature and flow rate (numbers $K^*_{T_g}$ and K_G). If Nu as a function of the above numbers is known, then the unsteady heat transfer coefficient may be found for any combination of performance parameters and boundary conditions.

Time variations of inlet temperatures and flow rates of hot and cold heat carriers are usually known when calculations are made of unsteady thermal processes in heat exchange apparatuses, including helical tube ones. A geometry of a heat exchanger is known, too.

In unsteady processes, the heat transfer coefficient is a nonlinear function of boundary conditions, i.e. a function not only of T_w/T_f but also of $\partial T_w/\partial \tau$, $\partial G/\partial \tau$, etc. Since even for the first heat exchanger cross-section from the inlet the values of T_w and $\partial T_w/\partial \tau$ are first unknown, the problem is being solved by the successive approximation method at each time step and along the apparatus. The accuracy of the method is infinitely high as the value of the time interval may be chosen as small as you wish.

The main point of the successive approximation method is as follows. The non-linear terms, i.e.

$$\alpha = \alpha_0(\tau) f(T_w, \partial T_w/\partial \tau) \tag{8.1}$$

are separated in the expression for the heat transfer coefficient. Here $f(T_w, \partial T_w/\partial \tau)$ is the function allowing for the effect of the boundary conditions on the heat transfer coefficient and $\alpha_0(\tau)$ is the coefficient taking into consideration the effect of the remainder parameters on heat transfer.

On this time interval, as a zero approximation at the moment τ, take the wall temperature $T_{wi}^{(0)}$ determined, assuming that $\alpha = \alpha_0(\tau)$ according to formula (8.1). In this case, a time step is chosen so that the heat transfer coefficient $\alpha(\tau)$ would vary almost linearly at this step. A knowledge of $T_{wi}^{(0)}$ for the considered cross-section i yields the time derivative of the wall temperature

$$\left(\frac{\partial T_w}{\partial \tau}\right)_i (0) = \frac{T_{wi}^{(0)} - T_{w,i-1}^{(0)}}{\Delta \tau} \tag{8.2}$$

Knowing $T_{wi}^{(0)}$ and $(\partial T_w/\partial \tau)_i^{(0)}$ provides the function allowing for the impact of the boundary conditions on the heat transfer coefficient $f^0(T_w, \partial T_w/\partial \tau)$ and the heat transfer coefficient $\alpha_0^{(1)}$ to a first approximation. The wall temperature $T_{wi}^{(1)}$ and its derivative

$$\left(\frac{\partial T_w}{\partial \tau}\right)_w (1) = \frac{T_{wi}^{(1)} - T_{w,i-1}}{\Delta \tau} \tag{8.3}$$

are found in terms of the obtained value of $\alpha_i^{(1)}$ to a first approximation. The values of $T_{wi}^{(1)}$ and $(\partial T_w/\partial \tau)^{(1)}$ obtained to a first approximation are the initial ones for a second approximation, and a cycle of calculations is repeated. If a required accuracy in determining T_{wi} is prescribed, then calculation must be continued until

$$\frac{T_{wi}^{(j)} - T_{wi}^{(j-1)}}{T_w^{(j)}} \leq \Delta$$

where j is the number of the approximation and Δ is the assigned accuracy in determining T_{wi}.

In calculating unsteady operating conditions of heat exchangers, it should be taken into account that the wall temperature also varies in time and, hence, the wall either accumulates some amount of heat from a hot heat carrier and it is spent for wall heating or it removes extra heat to a cold heat carrier and it is spent for wall cooling. A system of one-dimensional equations consists of two energy equations for both heat carriers and the heat balance equation for a wall. This system may be represented as

$$G_h \frac{c_{ph}}{u_h} \frac{\partial T_h}{\partial \tau} + G_h c_{ph} \frac{\partial T_h}{\partial x} = \Pi_h \alpha_h (T_w - T_h) \tag{8.4}$$

$$G_x \frac{c_{pc}}{u_c} \frac{\partial T_c}{\partial \tau} + G_c c_{pc} \frac{\partial T_c}{\partial (L - x)} = \Pi_c \alpha_c (T_w - T_c) \tag{8.5}$$

$$(\rho c F)_w \frac{\partial T_w}{\partial \tau} = \Pi_h \alpha_h (T_h - T_w) + \Pi_c \alpha_c (T_c - T_w) \tag{8.6}$$

where G_h, G_c are the flow rates of hot and cold heat carriers; c_{ph}, c_{pc} are their heat capacities; T_h, T_c are the mean-mass temperatures of hot and cold heat carriers in the considered cross-section; u_h, u_c are the mean flow-rate velocities of heat carriers; Π_h, Π_c are the wetted perimeters; α_h, α_c are the heat transfer coefficients; T_w is the wall temperature; L is the apparatus length; F is the total cross-section of bundle tube walls. The system of equations (8.4)–(8.6) is written, assuming that the countercurrent flow takes place in a heat exchanger, the longitudinal coordinate being taken from the hot heat carrier efflux.

The system of equations (8.4)–(8.6) [26] was reduced to a dimensionless form and then to a system of integral equations.

8.2 CALCULATION OF UNSTEADY HEAT TRANSFER IN HELICAL TUBE BUNDLES WITH REGARD TO INTERCHANNEL MIXING

A procedure to calculate in-line helical-tube bundle heat exchangers with regard to interchannel mixing is similar to the one described in Section 8.1. The difference consists mainly in allowing for interchannel mixing of heat carrier due to nonuniform heat supply to the bundle cross-section. Let us consider the effect of this factor, first, on a steady process. Since the most effective heat transfer enhancement is achieved by using helical tubes, when the thermal resistance is limiting in the intertube space, let us analyze a gas, e.g. air as heat carrier flowing past a tube bundle. In this case, either a liquid or a gas is flowing inside the tubes at the higher heat transfer coefficient α_1 and at a higher temperature, and the wall heat flux is determined by the formula

$$q_w = \alpha_1 (T_{b1} - T_{w1}) \tag{8.7}$$

When the thickness of spiral oval tubes is $\delta_w \ll d_{eq}$, the effect of the channel curvature on heat transfer may be neglected, and the problem on heat transfer through a flat wall may be solved. Then from the heat flux continuity condition

$$q_w = \frac{\lambda}{\delta_w} (T_{w1} - T_{w2}) \tag{8.8}$$

and

$$q_w = \alpha_2 (T_{w2} - T_{sp2}) \tag{8.9}$$

where α_2 is the heat transfer coefficient in the intertube space of a heat exchanger. Otherwise,

$$q_w - \chi (T_{b1} - T_{sp2}) \tag{8.10}$$

where

$$\chi = \left(\frac{1}{\alpha_1} + \frac{\delta_w}{\lambda} + \frac{1}{\alpha_2} \right)^{-1}$$

is the heat transfer coefficient, T_{sp2} is the local heat carrier temperature in the intertube space. At $1/\alpha_2 \gg 1/\alpha_1$, i.e. the thermal resistance in the intertube space is much greater than that inside the tubes

$$\chi \approx \left(\frac{1}{\alpha_2} + \frac{\delta_w}{\lambda} \right)^{-1}$$

Knowing a mean-mass heat carrier temperature distribution inside helical tubes over the bundle radius $T_{b1} = f(x,r,\varphi)$, it is possible to calculate wall and heat exchanger temperatures by supplementing the system of equations (1.15)–(1.18) for flow of a homogenized medium with equations (8.8) and (8.9). As a result, together with a heat carrier temperature distribution over the heat exchanger cross-section in the intertube space, it is possible to determine a temperature distribution in helical tube walls T_{w2} and T_{w1} (in a solid phase). In this case, a volumetric heat release density is

$$q_v = q_w \frac{\Pi}{F} = \frac{\Pi (T_{b1} - T_{sp2})}{F \left[\frac{1}{\alpha_2} + \frac{\delta_w}{\lambda} + \frac{1}{\alpha_1} \right]} \tag{8.11}$$

where Π is the heated perimeter of a helical tube and is the function of the x, r, φ-coordinates as well as the prescribed temperature inside helical tubes $T_{b1} = T(x, r, \varphi)$. Since under these steady operating conditions the mean thermal power, Q, of the apparatus as well as the heat carrier flow rates is known, the mean heat flux density will be equal to

$$(q_w)_{mean} = \frac{Q}{\Pi N l} = \frac{G_1 c_p (T_{out1} - T_{in1})}{\Pi N l} \qquad (8.12)$$

where T_{out1} and T_{in1} are the mean-mixed heat carrier temperatures at the tube outlet and inlet. The power Q may be also expressed in terms of the heat transfer coefficient $(\alpha_1)_{mean}$ mean over the bundle cross-section:

$$Q = (\alpha_1)_{mean} [(T_{b1})_{mean} - (T_{w1})_{mean}] \Pi N l \qquad (8.13)$$

Then a relative value of $\bar{q}_w$ is equal to

$$\bar{q}_w = \frac{q_w}{(q_w)_{mean}} = \frac{\alpha_1}{(\alpha_1)_{mean}} \frac{T_{b1} - T_{w1}}{(T_{b1})_{mean} - (T_{w1})_{mean}} \qquad (8.14)$$

and depends on x, r, φ, and a volumetric heat release density is determined by the expression

$$q_v = (q_w)_{mean} \frac{\Pi}{F} \bar{q}_w \qquad (8.15)$$

In formulas (8.7), (8.9) and (8.11) the heat transfer coefficients α_2 and α_1 are calculated by the criterial relations found from experiment. In the intertube space of the apparatus at $Fr_M \geq 100$ and $Re > 1000 (1 + 3.6 \, Fr_M^{-0.357})^4 (T_w + T_{fb}/2T_{fb})$ and $T_w/T_f = 1-1.5$ the local heat transfer coefficient α_2 is determined by the formula

$$\alpha_2 = 0.023 \frac{\lambda_f \rho^{0.8} u^{0.8} Pr_f^{0.4}}{d_{eq}^{0.2} \mu_f^{0.8}} (1 + 3.6 Fr_M^{-0.357}) (T_w/T_f)^{-0.55} \qquad (8.16)$$

At $Fr_M < 100$ heat transfer enhances. In the transient flow region at $Re_f < 1000 (1 + 3.6 \, Fr^{-0.357})^4 (T_w + T_{fb}/2T_{fb})$ heat transfer is improved to a greater extent than the one in the turbulent flow region and is described by the relation

$$\alpha_2 = 83.5 \frac{\lambda_f \rho^n u^n Pr_f^{0.4}}{d_{eq}^{1-n} \mu_f^n} Fr_M^{-1.2} (1 + 3.6 Fr^{-0.357}) \left(\frac{T_w}{T_f}\right)^{-0.55} \qquad (8.17)$$

where at $Fr_M < 924$ $n = 0.212 \, Fr_M^{0.194}$ and at $Fr_M \geq 924$ $n = 0.8$. At the starting length of a tube bundle for $x/d_{eq} = 3.75-14$ local developed turbulent heat transfer is described by the formula

$$\alpha_2 = 0.049 \left(\frac{x}{d_{eq}}\right)^{0.287} \frac{\lambda_f \rho^{0.8} u^{0.8} Pr_f^{0.4}}{d_{eq}^{0.2} \mu_f^{0.8}} (1 + 3.6 Fr_M^{-0.357}) \left(\frac{T_w}{T_f}\right)^{-0.55} \qquad (8.18)$$

and in the transient flow region, by the relation

$$\alpha_2 = 178 \left(\frac{x}{d_{eq}}\right)^{-0.287} \frac{\lambda_f \rho^n u^n \mathrm{Pr}_f^{0.4}}{d_{eq}^{1-n} \mu_f^n} \mathrm{Fr}_M^{-1.2} (1 + 3.6\mathrm{Fr}_M^{-0.357}) \left(\frac{T_w}{T_f}\right)^{-0.55} \quad (8.19)$$

The hydraulic resistance coefficient, ξ_2, necessary to close the system of equations (1.15)–(1.18) is determined at $\mathrm{Fr}_M \geq 100$, $\mathrm{Re}_f = 3 \cdot 10^3 – 5 \cdot 10^4$, $T_w/T_f = 1–1.42$ and $x/d_{eq} \geq 3.74$ by the formula

$$\xi_2 = 0.3164 \mathrm{Re}_f^{-0.25} (1 + 3.6\mathrm{Fr}_M^{-0.357}) \quad (8.20)$$

The relations in form (4.104) and (4.109) may be applied for a nonuniform heat supply field. In the case of flow inside the tubes at $S/d = 6.2–12.2$, $\mathrm{Re}_f = 7 \cdot 10^3 – 2 \cdot 10^5$, $T_w/T_f = 1–1.9$ and $\mathrm{Re}_f \times (d_{eq}/D)^2 = 1 \cdot 5–500$, where D is the curvature diameter of the oval channel symmetry line (for $S/d = 6.2$ $D/d_{eq} = 18.26$, for $S/d = 12.2$ $D/d_{eq} = 79.56$) the coefficients α_1, ξ_1 are determined by the relations [39]:

$$\alpha_1 = 0.072 \frac{\lambda_f \rho^{0.76} u^{0.76} \mathrm{Pr}_f^{0.4}}{d_{eq}^{0.24} \mu_f^{0.76}} \left(\frac{T_w}{T_f}\right) n \left(\frac{d_{eq}}{D}\right)^{0.16} \quad (8.21)$$

where n at $x/d_{eq} > 30$ is determined by the formula

$$n = -0.17 - 0.27 \cdot 10^{-5} (x/d_{eq})^{1.37} [(S/d_{eq})^{2.1} - 109.6] \quad (8.22)$$

$$\xi_1 = \frac{0.71}{\mathrm{Re}_f^{0.2}} \left(\frac{d_{eq}}{D}\right)^{0.27} - 0.008 \left(\frac{d_{eq}}{D}\right)^{0.67} \quad (8.23)$$

The quantity D/d_{eq} in (8.23) is determined from the expression [39]:

$$D/d_{eq} = 0.5 + \frac{8}{\pi^2}\left(\frac{S}{2d_{eq}}\right)^2 \quad (8.24)$$

The relations presented in Section 4.1 are employed to find stationary values of the effective turbulent diffusion coefficients K. These empirical transfer coefficients close the system of equations (1.15)–(1.18).

In calculations of unsteady heat transfer in helical-tube bundle heat exchangers with regard to interchannel mixing of heat carrier, one should consider the properties of the liquid flowing inside the helical tubes to solve the system of equations (5.17)–(5.21) under boundary conditions (5.22)–(5.24) describing the flow of a homogenized medium.

The unsteady dimensionless effective turbulent diffusion coefficient K_{uns} used to close a system of equations and determining a rate of heat and mass transfer in a bundle and, hence, interchannel mixing of heat carrier is calculated by the experimental relations found in Chapter 5.

The proposed approach to solve the problems on unsteady heat and mass transfer in heat exchange apparatuses and in helical-tube bundle heat exchangers may be adopted to calculate temperature distributions of a gas heat carrier in the intertube space of the apparatus and helical tubes (solid phase), which is of special importance for the equipment operating at high temperature levels and high heat flux densities.

REFERENCES

1. Alifanov, O. M. Identifikatsiya protsessov teploobmena letatel'nykh apparatov (Identification of heat transfer processes in vehicles). Moscow, Mashinostroyeniye, 1979, 216 pp.
2. Bukreyev, V. I. and Shakhin, V. M. Eksperimental'noye issledovaniye turbulentnovo neustanovivshevosya techeniya v krugloy trube (Experimental study of turbulent unsteady flow in a round tube). Aeromekhanika, Moscow, Nauka, 1976, pp. 180–187.
3. Vilemas, Yu. V., Dzyubenko, B. V., and Sakalauskas, A. V. Issledovaniye struktury potoka v teploobmennikakh s vintoobrazno zakruchennymi trubami (Study of the flow structure in helical tube heat exchangers). Izv. AN SSSR, Energetika i Transport, 1980, No. 4, pp. 135–144.
4. Subbotin, V. I., Ibragimov, M. Kh., and Ushakov, P. A., et al. Gidrodinamika i teploobmen v atomnykh energeticheskikh ustanovkakh (Hydrodynamics and heat transfer in atomic power plants). Moscow, Atomizdat, 1975, 408 pp.
5. Glikman, B. F. Nestatsionarnyye techeniya v pnevmogidravlicheskikh tsepyakh (Unsteady flows in pneumohydraulic ducts). Moscow, Mashinostroyeniye, 1979, 256 pp.
6. Godunov, S. K. and Ryaben'ky, V. S. Raznostnyye skhemy (Difference schemes). Moscow, Nauka, 1977, 440 pp.
7. Gulin, N. P., Raneyskaya, G. V., and Inozemtseva, L. I. Dinamika techeniya zhidkosti i koeffitsiyent soprotivleniya treniya pri oporozhnenii truboprovodov (Dynamics of liquid flow and friction resistance coefficient at pipeline discharge). Preprint FEI-1601, Obninsk, 1984, 18 pp.
8. Dzyubenko, B. V., Ashmantas, L. A., and Bagdonavichyus, A. B. Nestatsionarnoye peremeshivaniye teplonositeliya v puchkakh vitykh trub. Sovremennyye problemy gidrodinamiki i teploobmena v elementakh energeticheskikh ustanovok i kriogennoy tekhnike (Unsteady heat carrier mixing in helical tube bundles. Modern problems of hydrodynamics and heat transfer in the elements of energy installations in cryogenic technique). All-Union Correspondence Mechanical Engineering Institute Press, 1985, Vol. 14, pp. 9–14.
9. Dzyubenko, B. V., Vilemas, Yu. V., and Ashmantas, L. A. Peremeshivaniye teplonositelya v teploobmennike s zakrutkoy potoka (Heat carrier mixing in a swirled flow heat exchanger). J. Eng. Phys. (Russian), 1981, Vol. 40, No. 5, pp. 773–779.
10. Dzyubenko, B. V. and Ievlev, V. M. Teploobmen i gidravlicheskoye soprotivleniye v mezhtrubnom prostranstve teploobmennika s zakrutkoy potoka (Heat transfer and hy-

draulic resistance in the intertube space of a heat exchanger with flow swirling). Izv. AN SSSR, Energetika i Transport, 1980, No. 5, pp. 117–125.
11. Dzyubenko, B. V. and Sakalauskas, A. V. Vikhrevaya struktura potoka v teploobmennike s vitymi trubami (The vortex flow structure in a helical tube heat exchanger). Ibid., 1986, No. 3, pp. 151–157.
12. Dzyubenko, B. V., Sakalauskas, A. V., and Vilemas, Yu. V. Energeticheskiye spektry turbulentnosti v teploobmennike s zakrutkoy potoka (Energy turbulence spectra in a swirled flow heat exchanger). Ibid., 1983, No. 4, pp. 125–133.
13. Dzyubenko, B. V., Urbonas, P. A., and Ashmantas, L. A. Mezhkanal'noye peremeshivaniye teplonositelya v puchke vitykh trub (Interchannel mixing of a heat carrier in a helical tube bundle). J. Eng. Phys. (Russian), 1983, Vol. 45, No. 1, pp. 26–32.
14. Žukauskas, A. A. Konvektivnyy perenos v teploobmennikakh (High-performance single-phase heat exchangers). New York, Hemisphere Publishing Corporation, 1989, 515 pp.
15. Ievlev, V. M. Turbulentnoye dvizheniye vysokotemperaturnykh sploshnykh sred (Turbulent motion of high-temperature continua). Moscow, Nauka, 1975, 256 pp.
16. Ievlev, V. M., Dzyubenko, B. V., and Segal, M. D. Teplomassoobmen v teploobmennike s zakrutkoy potoka (Heat and mass transfer in a swirled flow heat exchanger). Izv. AN SSSR, Energetika i Transport, 1981, No. 5, pp. 104–112.
17. Zhukov, A. V., Sviridenko, E. Ya., and Matyukhin, N. M. et al. Issledovaniye mezhkanal'novo peremeshivaniya v reshetkakh sterzhney s malymi otnositel'nymi shagami i obobshcheniye fakticheskovo materiala dlya sistem s distantsioniruyushchimi provolochnymi navivkami (Study of interchannel mixing in rod grids with small relative pitches and generalization of the data on the systems with remote wire tube arrangement). Obninsk, Preprint FEI-799, 1977, 14 pp.
18. Zhukov, A. V., Sviridenko, E. Ya., and Matyukhin, N. M. et al. Issledovaniye gidrodinamiki slozhnovo techeniya v sborkakh sterzhney s distantsioniruyushchey provolochnoy navivkoy (Study of hydrodynamics of complex flow in assemblies of rods with remote wire tube arrangement). Obninsk, Preprint FEI-867, 1978, 17 pp.
19. Kalinin, E. K., Dreitser, G. A., and Yarkho, S. A. Intensifikatsiya teploobmena v kanalakh (Heat transfer enhancement in channels). Moscow, Mashinostroyeniye, 1981, 205 pp.
20. Koshkin, V. K. and Kalinin, E. K. Teploobmennyye apparaty i teplonositeli (Heat exchange apparatus and heat carriers). Moscow, Mashinostroyeniye, 1971, 200 pp.
21. Kramerov, A. Ya. and Shevelev, Ya. V. Inzhenernyye raschety yadernykh reaktorov (Engineering calculations of nuclear reactors). Moscow, Atomizdat, 1984, 736 pp.
22. Kutateladze, S. S. Pristennaya turbulentnost' (Wall turbulence). Novosibirsk, Nauka, Siberian Branch, 1973, 228 pp.
23. Markov, S. B. Eksperimental'noye issledovaniye skorostnoy struktury i gidravlicheskikh soprotivleniy v neustanovivshikhsya napornykh turbulentnykh potokakh (Experimental study of the velocity structure and hydraulic resistance in unsteady confined turbulent flows). Izv. AN SSSR, Mekh. Zhidkosti i Gaza, 1973, No. 2, pp. 65–74.
24. Kalinin. E. K., Dreitser, G. A., and Kostyuk, V. V. et al. Metody rascheta sopryazhennykh zadach teploobmena (Methods of calculating conjugated heat transfer problems). Moscow, Mashinostroyeniye, 1983, 232 pp.
25. Minsky, E. M. and Fomichev, M. S. O puti smesheniya Prandtlya, kriteriyakh i masshtabakh turbulentnosti (On the mixing Prandtl length, turbulence numbers, and scales). Izv. AN SSSR, Energetika i Transport, 1972, No. 6, pp. 119–123.
26. Koshkin, V. K., Kalinin, E. K., and Dreitser, G. A. et al. Nestatsionarnyy teploobmen (Unsteady heat transfer). Moscow, Mashinostroyeniye, 1973, 328 pp.
27. Dzyubenko, B. V., Segal, M. D., and Ashmantas, L. A. et al. Nestatsionarnoye peremeshivaniye teplonositelya v teploobmennike s vitymi trubami (Unsteady heat carrier mixing in a helical tube heat exchanger). Izv. AN SSSR, Energetika i Transport, 1983, No. 3, pp. 125–133.

28. Ievlev, V. M., Dzyubenko, B. V., and Dreitser, G. A. et al. Nestatsionarnyy teplomassoobmen zakruchennykh potokov v kanalakh slozhnoy formy (Unsteady heat and mass transfer from swirled flows in complex-geometry channels). Heat and Mass Transfer—VII, Minsk, HMTI Press, 1984, Vol. 1, Pt. 1, pp. 91–95.
29. Nikiforov, A. N., Fafurin, A. V., and Gerasimov, S. V. Issledovaniye skorostnoy struktury nestatsionarnykh turbulentnykh techeniy. Gazodinamika dvigateley letatel'nykh apparatov (Study of the velocity structure of unsteady turbulent flows. Gasdynamics of engines of the vehicles). Kazan, KAI Press, 1982, pp. 43–48.
30. Ainola, L. Ya., Koppel, T. A., and Lamp, Yu. Yu. et al. O koeffitsiyente treniya dlya laminarnykh nestatsionarnykh techeniy v trube (On the friction coefficient for laminar unsteady tube flows). Proc. Tallin Polytechnical Institute, 1981, No. 505, p. 315.
31. Panteleyev, A. A., Slesarev, V. A., and Trushin, V. A. Issledovaniye vliyaniya temperaturnoy nestatsionarnosti na teplootdachu v lopatkakh turbin (Study of the effect of temperature unsteadiness on heat transfer in turbine blades). Teploenergetika, 1981, No. 10, pp. 58–61.
32. Pelageychenko, K. I. and Segal, M. D. Metodika rascheta nestatsionarnykh teplogidravlicheskikh protsessov v poristoy srede s vnutrennimi istochnikami energovydeleniya (Methods of calculating unsteady thermal and hydraulic processes in a porous medium with internal heat-release sources). Preprint, I. Kurchatov IAE, No. 3581, 1982, 15 pp.
33. Popov, D. I. Nestatsionarnyye gidromekhanicheskiye protsessy (Unsteady hydromechanical processes). Moscow, Mashinostroyeniye, 1982, 240 pp.
34. Rozhdestvenskiy, B. L. and Yanenko, N. N. Sistemy kvazilineynykh uravneniy (Systems of quasi-linear equations). Moscow, Nauka, 1978, 330 pp.
35. Roache, P. J. Computational fluid dynamics. Hermosa Publishers, Albuquerque, New Mexico, 1976.
36. Rozanov, V. I. Metodika teplogidravlicheskovo rascheta puchkov sterzhney s navivkoy. Sovremennyye problemy gidrodinamiki i teploobmena v elementakh energeticheskikh ustanovok i kriogennoy tekhnike (Methods of thermal and hydraulic calculation of bundles of rods with tube arrangement. Modern problems of hydrodynamics and heat transfer in the elements of energy installations in cryogenic technique). All-Union Correspondence Mechanical Engineering Institute Press, 1985, Vol. 14, pp. 14–20.
37. Samarskiy, A. A. Vvedeniye v teoriyu raznostnykh skhem (An introduction to the theory of difference schemes). Moscow, Nauka, 1971, 552 pp.
38. Samarskiy, A. A. and Nikolayev, E. S. Metody resheniya setochnykh uravneniy (Methods of solving network equations). Ibid., 1978, 590 pp.
39. Danilov, Yu. I, Dzyubenko, B. V., and Dreitser, G. A. et al. Teploobmen i gidrodinamika v kanalakh slozhnoy formy (Heat transfer and hydrodynamics in complex geometry channels). Moscow, Mashinostroyeniye, 1986, 200 pp.
40. Ievlev, V. M., Danilov, Yu. I., and Dzyubenko, B. V. et al. Teploobmen i gidrodinamika zakruchennykh potokov v kanalakh slozhnoy formy (Heat transfer and hydrodynamics of swirled flows in complex geometry channels). Heat and Mass Transfer—VI, Minsk, HMTI Press, 1980, Vol. 1, Pt. 1, pp. 88–99.
41. Lokay, V. Ch., Bodynov, M. I., and Zhuykov, V. V. et al. Teploperedacha v okhlazhdayemykh detalyakh gazoturbinnykh dvigateley letatel'nykh apparatov (Heat transfer in cooled parts of gas-turbine engines of aircraft). Moscow, Mashinostroyeniye, 1985, 216 pp.
42. Teplofizika i gidrodinamika aktivnoy zony i parogeneratorov dlya bystrykh reaktorov (Thermophysics and hydrodynamics of an active zone and steam generators for fast reactors). Proc. of SMEA, Vol. 1, Prague, 1978, 209 pp.
43. Urbonas, P. A. Eksperimental'noye issledovaniye koeffitsiyenta gidravlicheskovo soprotivleniya v puchke vitykh trub. Sovremennyye problemy gidrodinamiki i teploobmena v elementakh energeticheskikh ustanovok i kriogennoy tekhnike (Experimental study of the hydraulic resistance coefficient for a helical tube bundle. Modern problems of hydrody-

namics and heat transfer in the elements of energy installations in cryogenic technique). All-Union Correpondence Mechanical Engineering Institute Press, 1982, Vol. 11, pp. 78–82.
44. Khabakhpasheva, E. M., Perepelitsa, B. V., and Pshenichnikov, Yu. M. Statisticheskiye kharakteristiki pul'satsiy temperatury v nestatsionarnom turbulentnom potoke (Statistical characterstitics of temperature fluctuations in the unsteady turbulent flow). Heat and Mass Transfer—VII, Minsk, HMTI Press, 1984, Vol. 1, Pt. 2, pp. 133–137.
45. Shkema, R. K., Drizhyus, M.-R. M., and Lappo, V. V. Inertsionnyye kharakteristiki sistemy "pnevmotrassa-datchik davleniya" (Inertia characteristics of the "pneumotic line-pressure pick up" system). Trudy AN Lit. SSR, Ser. B, 1984, Vol. 5 (144), pp. 79–84.
46. Shumakov, I. V. Metod posledovatel'nykh intervalov v termometrii nestatsionarnykh protsessov (Successive interval method as applied to the thermometry of unsteady processes). Moscow, Atomizdat, 1976, 216 pp.
47. Shchukin, V. K. and Khalatov, A. A. Teploobmen, massoobmen i gidrodinamika zakruchennykh potokov v osesimmetrichnykh kanalakh (Heat transfer, mass transfer, and hydrodynamics of swirled flows in axisymmetric channels). Moscow, Mashinostroyeniye, 1982, 200 pp.
48. Baumann, W. and Hoffman, H. Coolant cross-mixing of sodium flowing in line through multirod bundles with different spacer arrangements. Progress in Heat and Mass Transfer, New York, 1973, Vol. 7, pp. 114–120.
49. Baumann, W. and Moller, R. Experimental study of coolant cross-mixing in multirod bundles, consisting of unfinned, one, three, and six fin rods. ATKE, 1969, Vol. 14, pp. 298–304.
50. Collingham, R. E., Thorne, W. L., and Cormack, J. B. Coolant mixing in a fuel pin assembly utilizing helical wire-wrap spacers. Nucl. Eng. and Design, 1973, Vol. 24, No. 3, pp. 393–400.
51. Ievlev, V. M., Kalinin, E. K., and Danilov, Yu. I. et al. Heat transfer in the turbulent swirling flow in a channel of complex shape. Proc. 7th Int. Heat Transfer Conference, Munich, Hemisphere Publishing Corporation, 1982, General papers, Vol. 3, pp. 171–176.
52. Ievlev, V. M., Dzyubenko, B. V., and Dreitser, G. A. et al. In-line and cross-flow helical tube heat exchangers. Int. J. Heat and Mass Transfer, 1982, Vol. 25, No. 3, pp. 317–323.
53. Kawamura, H. Transient hydraulics and heat transfer in turbulent flow. Nuclear Technology, 1976, Vol. 30, No. 3, pp. 246–255.
54. Ohmi, M, Iguchi, M., and Akao, F. Laminar-turbulent transition and velocity profiles for oscillatory rectangular duct flows. Trans. Jap. Soc. Chem. Eng., 1983, Vol. 49, No. 447, pp. 2343–2353.
55. Okomoto, Y., Hishida, M., and Akino, K. Hydraulic performance in rod bundles of fast reactor fuels—pressure drop, vibration, and mixing coefficient. Proc. Symp. Progress in Sodium-Cooled Fast Reactor Engineering, Monaco, JAFA SM-130/38, March, 1970, 12 pp.
56. Skok, J. Mixing of the fluid due to the helicoidal wire of fuel pins in a triangular array. International Heat Transfer Seminar, Trogir, Yugoslavia, September, 1971, 10 pp.
57. Sutherland, W. A. Experimental heat transfer in rod bundles. Heat Transfer in Rod Bundles, ASME, New York, 1968, pp. 104–138.
58. Tu, S. W. and Ramaprian, B. R. Fully developed periodic turbulent pipe flow. J. Fluid Mechanics, 1983, Vol. 137, pp. 31–81.

INDEX

Acceleration, 37
Adiabatic air flow, 96
Aluminum oxide, 50
Anisotropy of properties, 71
Asbestos, 177
Aviation engineering, 4
Axial flow regions, 107
Azimuthal periodicity, 129
Azimuthal transfer, 3, 37, 105
Azimuthal velocity vector, 9

Bessel function, 96
Bundle:
    axis, 3, 11, 37
    boundaries, 129
    cell(s), 36, 110, 138
    cross section, 31, 36, 70, 115, 178
    diameter scale, 146
    inlet, 13, 48, 125, 129
    length, 13
    outlet, 142
    porosity, 38, 50
    radius, 7, 37, 47, 93
    shell diameter, 41

CAMAK crate, 59
Central rods, 112
Centrifugal forces, 33, 37, 108
Channel cross section, 172, 173
Channel outlet, 175
Channel wall, 6
Chromel-alumel thermocouple(s), 60, 100
Close-packed tube bundle, 175
Coarse vortices, 69, 71
Cold gas, 24
Connecting channels, 3
Convective transfers, 146
Cooper tips, 48
Core fluctuations, 26
Corrosion-resistant steel, 78
Corrosion-resistant steel shell, 50
Crossflow helical tube exchangers, 2, 3
Cross mixing of flow, 115
Curvilinearity, 44
Cylindrical wall, 167

Data processing, 184
Decreasing gas flow rate, 193
Decreasing heat load, 152
Diaphragms, 48
Digital voltmeter, 54
Dimensionless flow temperature, 138
Dimensionless velocity, 110
Displacement thickness, 7, 18, 93, 118
Dissipation, 16
Disturbance waves, 32
Drop liquids, 30

222 INDEX

Electric-conducting varnish, 116
Electric current, 115
Electric current busbars, 48
Electric fields, 55
Electric insulation, 45
Electric resistance, 170, 181
Electromagnetic gate valve, 61
Electronic pressure transducer, 55
Energy vortices, 68
Euler number, 35

Filter, 175
Finned rod(s), 83
Finned rod bundles, 4, 86
Fixed diffusion source, 44
Fixed solid phase, 125
Flat channel height, 76
Flow:
    acceleration, 26, 27, 31, 73, 75, 159, 160, 197
    core, 35, 43, 45, 66, 70, 71, 103, 109, 146, 153, 194
    deceleration, 27, 31, 75, 158, 160
    nonisothermicity, 29
    parameters, 11
    rate pulsations, 77
    swirling, 2, 4, 33, 72, 107
    turbulization, 2, 3, 72, 118
    velocity, 37
FORTRAN, 11, 124
Fourier number, 38, 133, 154, 155
    transformation, 68
Friction, 7, 19, 125
Froude number, 37
Frozen turbulence, 66

Galvanic conduction, 56
Galvanometer circuit, 181
Galvanometer damping, 183
Gas channel flow, 138
Gasdynamic disturbances, 158
Gasdynamic parameters, 33
Gasynamics, 32
Gas:
    flow, 24, 121, 180
    pressure, 22, 25, 191
    properties, 21
    velocities, 33, 174
Gauss distribution, 42, 43, 101
Glass-fibrous cloth, 47
Generator terminals, 48
Gravity, 37

Heat:
    capacity, 13
    carrier flow rate, 20, 156, 157, 161, 162
        time variation, 161
    carrier mixing, 41, 47
    carrier temperature, 11, 60, 81
        velocity, 3, 11
    conduction, 6
    diffusion, 41, 45, 52
    exchanger shell, 11
    flux density, 2, 20, 167, 175, 187
    leakages, 167, 171, 177
    load, 141
    power, 46, 52, 130, 156
    propagation velocity, 79
    -resistant glass fiber insulation, 179
    resistant varnish, 50
    transfer coefficient, 7
        enhancement, 2
Helical tube(s), 2, 3, 142
    axis, 10
    bundle heat exchangers, 1, 2
    tube bundle(s), 1, 9, 10, 16, 18, 32, 59, 65, 72, 93, 136, 159
    walls, 6
Hexagonal casing, 78
Hexagonal shell, 48
High-speed unsteadiness, 197
Homogeneous liquid, 5
Homogeneous turbulence, 42, 66
Hot air flow rate, 101
Hot gas flow rate, 194
Hot gas masses, 24
Hot junctions, 179
Hot-wire anemometer, 67, 72
Hydraulic diameter, 128
Hydraulic resistance, 3, 8, 17, 19, 26, 35, 37, 72, 98, 108, 114, 115, 116

Increasing heat load, 146
Inlet flow temperature, 175
Interacting spiral flows, 109
Interchannel mixing, 37, 88, 135, 211
Internal heat source density, 166
Internal heat release, 187
Intertube space, 1–3, 99
Inverse heat conduction, 166
Isotropic structure, 5, 69
Isotropic turbulence, 42, 66
Iteration cycles, 10, 14, 126

Jet outlet, 44

INDEX **223**

Karman's number, 35
Kinematic turbulent energy, 26
Kinematic viscosity, 128

Lagrange representation, 42
    of turbulent field, 40
Laminal flow, 5, 37
Lateral inlet, 3
Lewis and Prandtl numbers, 9
Liquid:
    circular channel flow, 111
    flow rate pulsations, 77
    heating, 195
    masses, 35
    -metal heat carriers, 35
Longitudinal heat carrier flow, 167
Longitudinal pulsation velocity, 65, 75
Longitudinal turbulent diffusion, 13
Longitudinal velocity vector, 127

Manometer, 48
Mass forces, 74
Mass transfer, 140, 156
    enhancement, 65
    processes, 1, 31, 52
Matrix factorization, 10
Mean-mass flow temperature distribution, 172
Measuring systems and apparatuses, 179
Mercury DF150, 180
Metal consumption, 3
Mica glass-ceramic plates, 176
Microcycle, 56
Mixing, 52
Molecular thermal conductivity, 138

Nikuradse's formula, 16
Nomograms, 19
Non-cooled surface radius, 79
Nonisothermal channel flows, 76
Nonuniform heat-releasing field, 96, 140
Nozzle, 61
Nozzle outlet, 48
Nusselt number, 20, 28, 112, 186

Ohmic resistance, 45, 175
Organic-glass inserts, 52
Orifice flow meter, 175
Oscillograph, 181
Outlet bundle temperature, 48
Outer tube wall temperature, 167

Oval helical tubes, 2, 83, 88
Oval helical tube heat exchangers, 1

Peripheral rods, 112
Peripheral tube row, 112
Pipe, 79
Pitot tube(s), 50, 93
Plate, 79
Poronite gaskets, 176
Porosity, 3, 13, 37, 38, 128
Prandtl's formula, 16
Pressure:
    drops, 11
    gauges, 180
    gradient, 10
    pulsations, 35
    regulator, 175
    transducer, 54
Probes, 54
Pulsational characteristics, 35
Pumping power, 1

Quasi-solid rotation law, 37, 107
Quasi-stationary heat transfer, 7

Radial velocity, 2
Radiative heat transfer, 59
Reed relays, 56
Reichardt's formula, 16, 21
Reynolds number(s), 5, 27, 33, 36, 71, 103,
    147, 156, 168, 180, 203
Rod, 79
Rod grid pitch, 84
Round tube(s), 37, 38, 75, 100, 114, 162

Shear stress(es), 26, 74, 109
Slotted channels, 50
Small-amplitude velocity pulsations, 23
Smooth tube, 3
Solid phase, 12, 31, 121
Solid-state law, 2, 134
Sonic velocity, 124
Sonic wave propagation, 121
Space temperature variations, 45
Spatial flow nonuniformity, 5
Spatial turbulence scale, 72
Spatial turbulence scale ratios, 91
Spectral density, 67, 68
Spiral:
    bundle channels, 37
    channels, 9, 35, 134

Spiral (*Cont.*):
    -finned rods, 98
    tube channel, 15
    tube surfaces, 118
Stagnation zones, 37
Static pressure, 52
Static pressure holes, 50, 180
Steady heat transfer, 30
Steady-state flow structure, 65
Steady temperature fields, 39
Steel hexagonal casing, 176
Straight helical tube bundles, 83
Strain gauges, 55
Stroke pulses, 58
Subsonic velocity, 32
Swirled flow heat exchanger, 2

Temperature:
    fields, 8, 47, 89, 154
    gauges, 59
    range, 29
Thermal:
    conductivity, 9, 13, 80
    diffusivity, 20, 79
    homogeneity, 133
    hydrodynamic wall layers, 37
    insulation, 166
    resistance, 135, 159
    stresses, 19
    unsteadiness, 32
    wake, 85
    wall layers, 37
Thermocouple(s), 44, 47, 51, 130
Thermocouple bend, 59
Thermophysical properties of heat carrier, 11
Tracer gas, 44
Transport properties of flow, 83
Transverse velocity, 7, 13, 33
Tube:
    bundle cells, 137
    bundle diameter, 13
    curvature radius, 37
    damp, 61
    oval, 93
    perimeter, 3
    plates, 1, 175
    walls, 45
    wall temperature, 150, 190
    wall thickness, 122, 190
Turbo-compressor, 47
Turbulence anisotropy, 44
Turbulence generation, 24, 28, 146
    intensity, 44
    level, 36

Turbulent:
    channel flow, 135
    diffusion coefficient(s), 12, 121
    energy, 23
    field, 42
    flow, 5, 31, 42
    gas flow, 22
    heat flux density, 15
    pulsation, 75
    shear stress, 15
    thermal conductivity, 193
    transfer characteristics, 17
    viscosity, 26
    viscosity coefficient, 26
    water flow, 77
Turbulization ratio, 35
Twisted bundle(s), 44
    of helical tubes, 53
Twisted helical tubes, 11
Twisting pitches, 2
Twisting tube pitches, 1, 3

Uniform heat supply, 108
Universal logarithmic profile, 18
Unsteady:
    flow temperature variation, 161
    heat, 1, 4, 15, 18, 21, 23, 25
        and mass transfer process, 31, 52, 209
    heat transfer in helical tube bundles, 165, 189, 198
    temperature fields, 78, 121
    turbulent flow structure, 73

Van Driest's formula, 16
Velocity, 3
    gradient, 26
    profile development, 7
    pulsations, 23, 44, 68
    vector, 7, 34, 124
Vibration strength, 1
Viscosity, 30
    coefficient, 6, 8
    forces, 15
Viscous sublayer, 23
Viscous sublayer thickness, 77
Vitoshinsky's nozzle, 48
Voltage drop, 179
Voltage polarity, 54
Volumetric energy release, 93
Vortex:
    flow structure, 65
    motion, 9

Vortex (*Cont.*):
   rotation, 3
   structure(s), 23, 71
   transfer, 134, 146

Wake, 36
Wall:
   curvature radius, 15
   flow region, 38
   flux density law, 29
   heating, 195
   layer thickness, 5, 15, 119
   layer turbulization, 146
   temperature, 139, 155, 199
   thickness, 21, 167
Water gauges, 180
Wave numbers, 68
Wetted perimeters, 13
Wire rods, 4